Plant Responses to Environmental Stimuli

Michel Thellier

Plant Responses to Environmental Stimuli

The Role of Specific Forms of Plant Memory

Springer

éditions Quæ
Éditions Cirad, Ifremer, Inra, Irstea
www.quae.com

Michel Thellier
Emeritus Professor of the University
of Rouen
Rouen, France

ISBN 978-94-024-1475-2 ISBN 978-94-024-1047-1 (eBook)
DOI 10.1007/978-94-024-1047-1

Éditions Quæ, R10, 78026 Versailles cedex, France www.quae.com

Printed on acid-free paper

This Springer imprint is published by Springer Nature
The registered company is Springer Science+Business Media B.V.
The registered company address is: Van Godewijckstraat 30, 3311 GX Dordrecht, The Netherlands

Foreword

Yes, plants possess memory ability! In his little book, Michel Thellier has combined recent discoveries with older data dealing with plant memory and its potential role on plant acclimatization to environment stimuli. By placing memory within an evolutionary frame, the author persuades us that a new way of researches has opened in plant physiology.

This programme appears as a new step in an already long (though being badly known) story. The existence of memory implies the ability to perceive signals and therefore some sensitivity. Strangely enough, the scientific reflexion about plant sensitivity has been occulted though a number of philosophers and botanists have taken it into consideration. Strangely also, the simplistic Aristoteles' ideas continue being cited, although they were almost immediately disproven. The first biologists have taken interest in the differences existing between animals and plants. According to Aristoteles (384–322 before Christ), a soul was present in each living being, but with different features in the different beings. The human soul had three functions: vegetative, sensitive and mental. Only the two firsts were present in animals, while plants possessed only a vegetative soul governing generation, feeding and growth; in other terms, plants were devoid of sensitivity, a fortiori of memory. Theophrastus (371–288 before Christ), who was interested by plant phototropism, criticized this oversimple Aristoteles' viewpoint.

Since the very beginning of modern science, the plant mystery has been the object of a lot of discussions, though, obviously, neither the concepts nor the techniques requisite for a rational approach were available at that time. Several key observations occurring during Renaissance are reported; however, in most cases, their interpretation was wrong. Both Giambattista della Porta (1535–1615), who is better renowned for his researches in optics, and William Gilbert (1544–1603), a pioneer about modern electricity, have taken interest in plant polarity; using an audacious comparison with magnet behaviour, they have explored the way towards a mechanistic explanation of plant movements. The Portuguese botanist Cristobal Acosta (1515–1594) was interested in the rapid movements of the leaves of the sensitive plant (*Mimosa pudica*) and in the circadian movements of the leaves of tamarind (*Tamarindus indica*) that coil in the evening and uncoil in the morning.

Francis Bacon (1561–1626) believed that these movements were a mere result from mechanical actions; he proposed the plant vessels and fibres to bear a strong analogy with the nerves and blood vessels of animals. William Harvey (1578–1657), who discovered blood circulation and the mechanical role of the heart, has put forward that the cause of the rapid movements of the leaves of the sensitive plant is similar to that of muscle contraction in animals.

Surprisingly, the question of plant sensitivity did not remain restricted to a few intellectuals: in the middle of the seventeenth century, the general public developed a true infatuation for plant movements. In Great Britain, sensitive plants were extensively grown in greenhouses. Moreover, experimenters like Robert Browne (1605–1682) and Henry Power (1623–1682) began to study phototropism, gravitropism and nutation. King Charles II (1630–1685) even asked the Royal Society to find an explanation to the movements of the leaves of the sensitive plant. The experiments that followed, as reported by Robert Hooke (1635–1703), were devoted to plant sensitivity. All that has supported the idea that, in the great chain of being (the so-called *scala naturae*), the sensitive plant was an intermediate between plant and animal, as this was already the case with the soft-water hydra (*Hydra viridissima*) that was discovered by Abraham Trembley (1710–1784). As a consequence, it was no longer possible to consider sensitivity, or "irritability" according to Charles Bonnet (1720–1793), belonging only to animals. In the eighteenth century, Julien Offray de la Mettrie (1709–1751), the author of the provocative book *L'Homme Machine* [The Machine Man] (1747), also wrote 1 year later *L'Homme Plante* [The Plant Man] in which he compared the irritability of these two organisms with each other.

With the generalization of experimental research, at the onset of the nineteenth century, several naturalists undertook a thorough study of plant sensitivity. As soon as 1806, Thomas Knight (1759–1838) experimented on gravitropism; for that, he invented the so-called Knight's wheel, demonstrating that the root and the aerial part of a plant orient according to the acceleration to which they are subjected (gravitation only or gravitation combined with a centrifuge force); in 1812, he also showed that the root exhibits a negative phototropism. Henri Dutrochet (1776–1847) suggested an interpretation for plant growth movements via osmotic processes; by longitudinally slicing a young plant in two parts and exposing one part to light and the other to darkness, he discovered that the illumined part bent, while the other one continued to grow straight on. The German Julius von Sachs (1832–1897) and the Austrian Julius von Wiesner (1838–1916) attempted to determine whether plant movements were active responses to the environment or purely passive consequences of the effects of light and gravity; they examined whether the movement quantity was proportional or not to the intensity of the stimulus; Sachs, pioneering a microscope approach, had a suspicion that the intracellular structure may depend on the environment.

At the same period of time, last but not least, the name of Charles Darwin (1809–1882) came out in the literature. Apart from his famous book *On the Origin of Species by Means of Natural Selection*, Darwin has written a number of publications dealing with botanic and plant physiology, in which the concepts of plant evolution and adaptation were close to emerging. In the year 1855, he studied the seed longevity in saltwater; clearly, he was wishing to test if plant colonization of the volcanic islands took place via the marine way. In the course of all his life, he was interested in plant reproduction, especially that of orchids and Fabaceae (which were termed Papilionaceae at that time), with particular reference to possible hybridizations and to the essential role of pollinating insects such as Hymenoptera or Lepidoptera. In the years 1860 and following, Darwin conceived a passion for plant movements and sensitivity. Two important books deal with climbing plants (*On the Movements and Habits of Climbing Plants*, 1865) and with insectivorous plants and their capture movements (*Insectivorous Plants*, 1875). He then urged his seventh child, Francis (1848–1925), to carry on broader observations and studies of plant movements; this was the matter of his last book (*The Power of Movements in Plants*, 1880). It is not a surprise to find in the book by Michel Thellier the same dramatic species as studied by Darwin, such as bryony and Venus flytrap.

Darwin has been especially interested in insectivorous plants. He has shown that a drosera (*Drosera rotundifolia*) was able to distinguish between different types of objects and had sensitivity better than that of human skin. The relatively rapid closure of the *Dionaea* traps caused Darwin to speculate on the possible existence of a nervous system in plants. In order to put that idea to the test, he visited the physician John Burdon-Sanderson (1828–1905) at the University of London in 1873. Burdon-Sanderson set electrodes on *Dionaea* leaves, and, each time he touched one of the hairs present in the *Dionaea* traps existing at the extremity of the plant leaves, he registered an electrical wave similar to the action potentials of animal neurons. Darwin related the result of that beautiful experiment on sensitivity, to the plant adaptation to a very peculiar mode of nutrition.

In 1871, St. George Mivart (1827–1900), a professor in London University, raised objections on several points of *On the Origin of Species*. In particular, Mivart stated that natural selection could not explain the way how plant sensitivity appeared, especially in the climbing plants. In his response, Darwin postulated that circumnutation was a universal property of plants and that it is thus an ancestral feature that has subsequently specialized to give the various other types of movements. This is one of the events that have caused Darwin to take gravitropism and phototropism into consideration. A series of experiments allowed him to assume that the root tips was sensitive to gravity, and to other sorts of stimuli such as touching and light, and that a signal was transmitted to the growth area causing the root to bend (the so-called *root cap hypothesis*). After a debate with Sachs, Darwin proposed a *root-brain hypothesis* that is interpreted as follows in the last pages of his book *The Power of Movements in Plants*: "it is hardly an exaggeration to say that the tip of the radicle thus endowed, and having the power of directing the movements of the adjoining parts, acts like the brain of the lower animals". Darwin did not propose the existence of some sort of memory, but he wasn't so far from it. Let us finally recall a series of experiments about the coleoptile of Poaceae, as quoted by Michel Thellier. Nowadays, the experiments by Peter Boysen Jensen (1883–1959) and Frits Went (1903–1990) are most often cited, but it shouldn't be forgotten that Darwin, with the help of his son, was the first to show that light stimuli were perceived at the level of the root tip and that a corresponding signal migrated towards the neighbouring tissues.

In the English-speaking universities, the plant movements were included up to 1935 in the books dealing with compared psychology. It is only later that the development of plant physiology finally permitted this branch of instruction to become autonomous, with its own biochemistry and hormonology. Though nowadays ranking plant tropisms within psychology would seem absurd, this unitary view of the living world had the advantage not to isolate plants from animals, which isolation risks to astonish the general public when facing the reality of plant sensitivity.

In most cases, plant sensitivity is responsible for non-reversible growth movements such as the circumnutating of voluble plants or the root and stalk tropisms. Darwin was well aware of the adaptive importance of these movements and had set it in an evolution frame. The book by Michel Thellier goes a step further. Plants do have memory; however plant memory is totally different from the animal memory that is familiar to us. If the movements resulting from plant sensitivity have a crucial role in plant life (the root towards ground and the stalk towards light), memory enables plants to produce a more elaborate and more integrated response than would be a straightforward response. No doubt that the book by Michel Thellier will give rise to subsequent works that, from cellular and molecular biology to ecophysiology, will contribute to the understanding of how the 369,000 or so recorded species of angiosperms (according to the recent report of Kew Garden) have settled our planet.

For More Details

Hopper SD, Lambers H (2009) Darwin as a plant scientist: a Southern Hemisphere perspective. Trends Plant Sci 14(8):421–435

Kutschera U, Briggs WR (2009) From Charles Darwin's botanical country-house studies to modern plant biology. Plant Biol 11:785–795

Kutschera U, Niklas KJ (2009) Evolutionary plant biology: Charles Darwin's forgotten synthesis. Naturwissenschaften 96:1339–1354

Whippo CW, Hangarter RP (2009) The "sensational" power of movement in plants: a Darwinian system for studying the evolution of behaviour. Am J Bot 96(12):2115–2127

Institut de Biologie Paris-Seine
UMR 7138, "Evolution Paris-Seine"
7 Quai Saint Bernard
F-75252 Paris Cedex 05, France
e-mail: herve.le_guyader@umpc.fr

Hervé Le Guyader

Acknowledgements

In the preparation of this book, I have benefited of the gracious help of colleagues, friends and relatives for the form and the substance. I thank them all heartedly. I am especially indebted to Anne Alexandre, Jean-Louis Bonnemain, Marie-José Costil, Claude Gillet, Michel Gounot, Janine Guespin, Chantal and Yannick Kerdudou, Philippe Lefrançois, Nolwenn Legrand, Hervé Le Guyader, Ulrich Lüttge, Victor Norris, Sandrine Pesnel, Jeanine Rens, Camille Ripoll, Marie-Claire Verdus and Alain Vian.

Contents

Plant and Recollection

Once upon a Time…

Once upon a time, there was a young man and a young woman who were in love with each other. As they were wandering in the forest, they came close to a deep hollow, the flanks of which were covered with flowering myosotis. The young woman admired the beautiful flowers. The young man, venturing on the edge of the yawning chasm, started to collect a bouquet for her. He had never seen such a chasm, and he was not aware of the danger of the thick mud that it contained. He slipped on the wet wall down to the bottom of the hollow and began to sink. In an ultimate effort, he threw the flowers that he was handling towards the young woman and said:

Forget me not!

Then he was engulfed forever. Nobody knows what the fate of the young woman was. Did she live henceforth in the memory of her lost sweetheart? Did she soon get over it in the arms of another lover? Be it as it may, the myosotis became a symbol for souvenir. It was called *forget-me-not* in Great Britain, *ne-m'oubliez-pas* in France, *vergissmeinnicht* in Germany, *nomeolvides* in Spain and so on in many other languages.

Is the above not more than a romantic tale, or is it possible that myosotis (and other plants) really have memory capacity? There is no general agreement about that. A few people are persuaded that their house plants possess a memory almost similar to ours, remember the tenderness or the carelessness with which they have been treated and react accordingly. Others think that memory ability exists only in animals, especially in humans, and that "plant memory" is an oxymoron. Can truth be midway between these two extreme opinions?

A Brief Introduction to the Book

Nowadays it is generally admitted that plants possess rudimentary forms of memory, extremely different though from our own memory. The history of this discovery and of the role of plant memory is rather tortuous. The whole story started with the study of the correlations that exist between the different organs of a plant and on the possibility to modify these correlations by administering adequate stimuli. It was found that some stimuli, which were without any visible effect on the moment, however caused, at a later time, metabolic or growth changes within the plants. At first sight, this seemed to be non-understandable, but we could eventually interpret it by assuming that plants had a capacity to "memorize" information and that the stored pieces of information could remain latent during a longer or shorter lapse of time before taking effect in the control of plant development.

Studying this unexpected memory was fascinating but complicated because manipulating the experimental system with which it had been discovered was intricate. We found however other experimental systems, the manipulation of which was much easier. Other groups, using very diverse systems, also evidenced manifestations of memory in plants. It remained however to understand what was the advantage for plants to possess memory. By contrast with animals, plants are unable to move around for searching a satisfactory environment. Hence they must succeed to make use of the favourable stimulations that they receive (rain, light, etc.) while resisting efficaciously to the deleterious ones (cold and drought stresses, attack by herbivores, etc.) at the place where they have rooted (Fig. 7.1). Hints have appeared suggesting that memory is involved in the plant accommodation to their environment conditions. This little book recounts the history of plant memory as the author has lived through. Though it is issued from laboratory studies, it is not written in the form of a scientific publication.

Scientists are unusual people. Professionally, they are trained to plan and interpret experiments, constantly going back and forth between experimentation and reflexion. They have created a terminology, symbols and mathematical formulas that enable them to communicate with one another concisely and precisely, but that are obscure to the general public. Here we are going to use the natural language as much as possible. The experiments will be described in a simplified manner (the references to the original scientific articles being however added for those keen on exactness). At the end of the book, a glossary will give the meaning of the less usual terms.

For describing the investigations relative to plant sensitivity and memory, we shall have to speak about proteins, genes, DNA, messenger RNA, etc. These terms are not familiar to everybody, especially in the coding problems that they involve. A first appendix will be devoted to a series of games consisting of coding and deciphering secret messages with the help of black and white circles, squares and diamonds. Once at ease with these games, the coding used by life will not be a difficulty any longer. A second appendix will give basic data of molecular biology that are sufficient to understand the main text. Finally a third appendix will introduce the concept of ionic condensation, a phenomenon often ignored by biologists though it may play a decisive role of switch in the functioning of biological systems.

Welcome to the plant world, with their sensitivity to environment stimuli and the memory effects induced by these stimuli.

Chapter 1
Me, a Plant

1.1 A Little Bit of Plant Morphology

The young plant being the hero of the book let us begin by briefly presenting its main features. Most often, it is in the nursery school that one discovers seed germination. The teacher lays a few seeds, for instance of radish (*Raphanus sativus*), on waterlogged cotton in a glass box. In the following days, the seeds inflate. Then a root appears that creeps downwards through the cotton fibres while the aerial parts progressively develop upwards.

The image of a radish plant, approximately 2–3 weeks old, is visible in Fig. 1.1. The root, the hypocotyl, the terminal bud (also termed apex), the first leaf and two opposite cotyledons are clearly visible. The elongation area of the root is close to the tip. The cotyledons, which have approximately the shape of a heart in radish, pre-exist in the seed and enlarge and become progressively green after germination. The hypocotyl is the part of the stem between cotyledons and root. The apex is an actively growing bud. At the axil of each cotyledon, there is a bud (termed "cotyledonary bud") that remains inactive as long as the terminal bud is not removed. When the terminal bud is present, the cotyledonary buds are so small that they are hardly visible to the naked eye.

The apex forms the stem and new leaves. The radish leaves are quasi elliptic and slightly indented on the edge. However, the shape of the leaves is extremely variable depending on the plant species. The arrangement of the leaves is also variable. They can be opposite (i.e. arranged by pairs in which the leaves are displayed symmetrically from each other) or alternate (i.e. shifted at a distance from one another along the stem). The leaves, including the cotyledons, are made of the limb (the flat part) and the petiole that attach the limb to the stem. The leaves contain chlorophyll, a pigment that confers to them their green colour and permits them to carry out photosynthesis.

In the medium in which it is planted, the root draws the water and the mineral salts (nitrates, sulphates, phosphates, potassium, calcium, magnesium,

M. Thellier, *Plant Responses to Environmental Stimuli*,
DOI 10.1007/978-94-024-1047-1_1

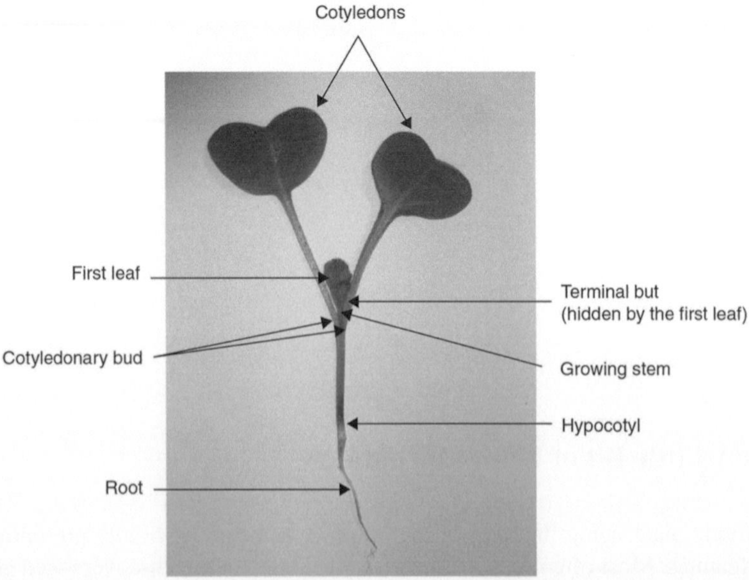

Fig. 1.1 Radish plant at the age of 2–3 weeks (Photo: Chantal Kerdudou). At this age, the hypocotyl, which will later be loaded with reserve substances and become the edible part of the plant, is already slightly pink. The terminal bud is actively growing and begins to form a stem and leaves. The cotyledonary buds are so small that they are not visible on the photo

oligo-elements, etc.) that the plant needs for its development. The absorption of mineral salts by the root is termed the mineral nutrition of the plant.

1.2 A Little Bit of Plant Physiology

The combustion of organic matter, including that originating from plants, causes consumption of oxygen, liberation of water and carbon dioxide (CO_2) and production of energy. Chemists know how to make the reverse reaction, i.e. creation of organic molecules by supplying energy to combine carbon dioxide with water and liberate oxygen. Photosynthesis enables the green plants to do the same, i.e. creation of organic matter (especially sugars), liberation of oxygen and consumption of carbon dioxide with light being the source of energy. Under natural conditions, the light absorbed by photosynthesis is simply that of the sun. Under laboratory conditions, light may be supplied by lamps.

Here we shall not describe the complex mechanism of photosynthesis. Simply, we shall recall that plants, by combining the sugars produced by photosynthesis with the mineral salts absorbed by the root, form the molecules constituting living matter. This is the way how green plants produce their own substance. Then herbivorous animals, fungi and all sorts of non-photosynthetic creatures feed on plants, both to elaborate their own substance and to produce the energy that they need. Finally, the carnivorous animals feed on the herbivores.

Trees, herbs, soft-water and marine algae and phytoplankton are the main photosynthetic beings that relieve us from carbon dioxide and produce the oxygen that we need for respiring. The animal life as we know it on earth wouldn't have existed in the absence of photosynthetic plants. It is noteworthy that a few animals exist that harbour photosynthetic systems in their own tissues and can thus live without feeding. For instance, the small sea worm *Convoluta roscoffensis* (recently renamed *Symsagittifera roscoffensis*), easily recognised due to its green colour, lacks a digestive system and lives on photosynthesis only. Are small green men, as encountered in science fiction stories, really existing photosynthetic humanoids? In the future, space exploration will possibly answer that question!

As long as humans were satisfied with feeding on organic matter or burning it (dead wood for instance), a quasi-stationary state was established, in which the consumption of oxygen and the production of carbon dioxide resulting from the activity of non-photosynthetic systems were compensated by the production of oxygen and fixation of carbon dioxide by the photosynthetic systems.[1] But when humans began to massively burn fossil combustibles (coal, fuel and gas), this quasi-stationary state was broken, and the concentration of carbon dioxide increased steadily, with the climatic consequences that may follow.

1.3 The Plant Cells

As all living systems, plants are made of cells. These cells are not as strictly specialised as animal cells are; however, their size and shape differ in the different tissues.

- The epidermis is a superficial tissue the cells of which are quasi rectangular and tightly joint with one another.
- At the surface of various organs, especially the leaves, stomata are made of cells forming an aperture that they can open (or close) to permit (or limit) the exchanges of gazes (especially water vapour). The image of a stomata inserted among the epidermal cells of the hypocotyl of a young flax plant is visible in Fig. 4.2.
- Parenchyma is an internal tissue made of usually rounded cells possessing chlorophyll and carrying out photosynthesis.
- There are two types of conductive tissues. First, the wood (or "xylem") vessels are made of dead cells constituting a tube with thick walls. These vessels convey passively the xylem sap (mainly made of water and mineral salts) from roots to leaves. The evaporation of water at the level of the leaves is the main process

[1] Photosynthesis is not the only existing mechanism for carbon dioxide sequestration. The mere dilution of the atmospheric carbon dioxide in the oceans and its immobilization in the form of calcium carbonate especially by shells and calcareous algae are also of importance.

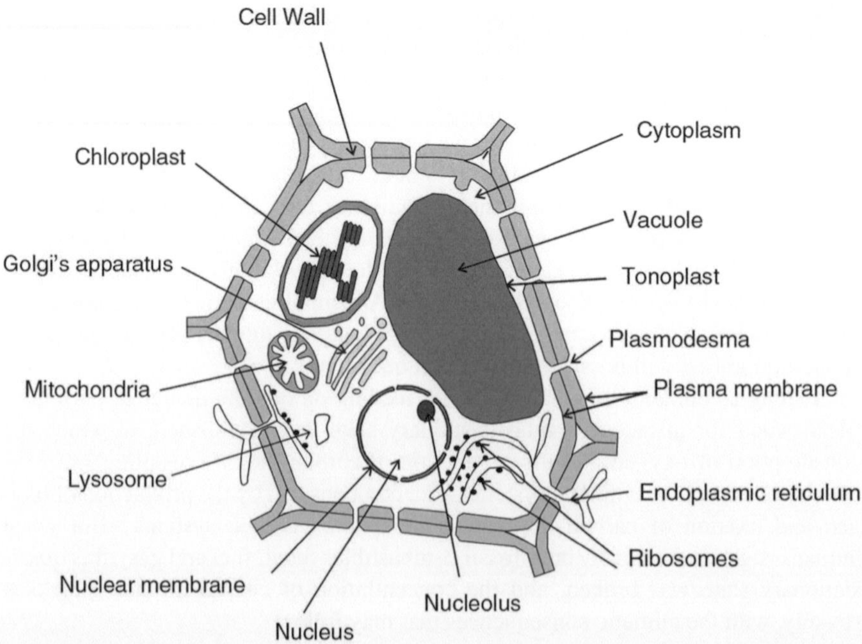

Fig. 1.2 Schematic representation of a plant cell (From Farineau and Morot-Gaudry 2011. Drawing by Y. Roux reproduced with his kind permission)

driving the ascension of the xylem sap. Second, the "phloem" cells are living cells that transport actively the elaborate sap (i.e. a sap enriched with organic substances formed in the leaves) from the leaves to the rest of the plant.

- By contrast with the tissues made of differentiated cells as described above, meristems are embryonic tissues made of small, non-differentiated cells. In the terminal bud (possibly also in other active buds in adult plants), and at the root tip as well, meristems exist in which cells are rapidly dividing and are responsible for increasing the lengths of the stems and roots. Such meristems are termed "primary meristems". Secondary meristems also exist for plant growth in diameter.

By contrast with animal cells, it is relatively frequent that differentiated plant cells dedifferentiate (under the effect of hormones, stresses, etc.) and begin dividing, thus producing new meristems. We are going to see in Sect. 4.3 a case in which young flax plants produce epidermal meristems in their hypocotyl under the effect of various stimuli.

Plant cells (Fig. 1.2) are compartmented systems in which the cell wall, the cytoplasm and one or several vacuoles are encountered from exterior to interior. The cell wall is mainly constituted of cellulose fibres embedded in a mixture of various sorts of pectin. The cytoplasm contains the nucleus (in which most of the genetic information is located) and various organelles such as mitochondria (the site for respiration), chloroplasts (the site for photosynthesis), endoplasmic reticulum and Golgi's apparatus (involved in the biosynthesis and migration of some proteins), lysosomes (involved in degradation processes) and a network of microfilaments and microtubules termed the cytoskeleton (not represented in Fig. 1.2). The nucleus contains one or several substructures termed the nucleoli. Mitochondria and chloroplasts contain some genetic information. Nucleus and organelles bathe in a complex aqueous medium (termed the "cytosol") that contains mineral ions, small organic molecules and macromolecules. Ion gradients are formed within the cytosol. A pseudo-compartmentation, caused by the accumulation of protein aggregates, exists in the cytosol. Vacuoles contain a nonstructured aqueous medium. Lipoprotein membranes border the main cellular compartments and organelles. For instance, the plasma membrane separates cytosol from cell wall and tonoplast separates cytosol from vacuoles. The cell wall is pierced with pores, termed plasmodesmas, which enable plant cells to exchange substances with one another.

Large variations take place with regard to this basic scheme. For instance, in meristem cells the wall is extremely thin, the vacuoles almost non-existing and the nucleus occupies a large part of the cell volume.

Chapter 2
Plant Sensitivity to Stimuli

2.1 Animal vs. Plant Sensitivity

There is no doubt that animals, especially humans, are sensitive beings. They perceive stimuli (whether visual, acoustic, tactile, olfactory, etc.) with their sensory organs, and they process corresponding information with their nerve cells (also termed neurons). By contrast, plants are often considered to be inert and insensitive. Isn't it said that somebody whose brain has been irreversibly damaged has become a vegetable for the rest of his life? In fact, there are stimuli that plants perceive and others that they do not. For instance, if plants may be sensitive to acoustic vibration, it does not seem that they have a special taste for classical music (Braam and Davis 1990), but they react to abiotic stimuli such as the relative length of night and day, gravitation, wind, rain, touching, drought, coldness, various sorts of vibrations and even electromagnetic radiations such as used in mobile telephones (this last point being discussed in more detail in Sect. 4.3.3). Plants also react to wounding (such as inflicted by herbivorous animals) and to attack by pests (fungi, bacteria or viruses).

Most often, the ultimate response of plants to stimuli is a modification of metabolism, growth and/or morphogenesis. In a few cases, the plant response involves a macroscopic movement (i.e. a movement visible to the naked eye). Sometimes, the response interests the whole plant and sometimes it is local (i.e. affecting only the organ that has perceived the stimulation) or at a distance from the stimulated area.

M. Thellier, *Plant Responses to Environmental Stimuli*,
DOI 10.1007/978-94-024-1047-1_2

2.2 The Various Sorts of Stimuli That a Plant Perceives

2.2.1 The Relative Length of Night and Day

In towns it may occur that a street lamp be very close to the branches of a tree. In that case, in autumn, when most of the leaves will have fallen, you may observe that the leaves closely neighbouring the lamp will still be there for a while (Fig. 2.1). The reason is that one of the stimuli causing leafs to fall is that nights have become long enough. This is the explanation why leaves fall in autumn, i.e. at a time when nights are longer. The lamp, by artificially decreasing the night length, prevents the leaves in its close vicinity to fall.

For some plant species (not all), the relative length of night and day (also termed "photoperiodicity") plays a major part in the induction to flowering. There are two types of plants exhibiting such sensitivity. The classical terms to characterize them

Fig. 2.1 An example of plant sensitivity to the relative length of night and day. This photo was taken at the onset of daylight on a slightly foggy autumn morning. All the leaves of the tree have already fallen except in the close vicinity of the lamp. By artificially shortening the length of the night period, the presence of the street lamp has retarded the fall of the leaves

are "short-day plants" and "long-day plants", meaning that flowering is induced when the daily period of daylight is short enough or long enough, respectively. However, here we shall preferably term them "long-night plants" and "short-night plants", because it is the length of the night, rather than that of the day, which is the important parameter. A long-night plant is not necessarily a plant requiring night to be longer than day. It is a plant that will be (or will not be) induced to flowering if the length of the night is above (or below) a given threshold. For instance, *Xanthium pennsylvanicum* is a long-night plant, the threshold of which is 9 h. This means that this plant will be induced to flowering if the daily period of night is above 9 h (the daily period of daylight thus being less than 15 h). Moreover, if night is interrupted by a short illumination (not more than a few minutes), the plant will not be induced to flowering. A plant that has not been induced to flowering will not flower. A plant that has been induced to flowering will actually flower if other requirements (e.g. concerning temperature, nutrition, photosynthetic activity) are fulfilled. The interested reader may have fun in trying to find out if some plants in his garden are likely to be long-night species (Box 2.1).

Annual plants germinate in spring and die in autumn. An annual, long-night plant, the threshold of which is close to the length of the night at the time when germination takes place, may have an unexpected behaviour. Indeed, germination

Box 2.1 The Forsythia

In my garden there is a foot of forsythia that I have observed during a full year. Forsythia is a woody perennial shrub that gets covered with beautiful yellow flowers in the late winter or early spring without forming a single leaf. Later on, the flowers wither and the forsythia comes into leaves. From that moment on, and during all summer and beginning of autumn, the forsythia remains covered with leaves without producing a single flower. In autumn the leaves fall, the plant enters in dormancy and only bare twigs will exist during the bad season. Yet when observing the plants carefully, one can see that these bare twigs produce one or a few flowers even in the depth of winter. At least, this is what has occurred with the variety of forsythia in my garden. In late winter or beginning of spring, the cycle starts again.

Although coming to an undisputable conclusion would require a more thorough study, all this is likely to mean that forsythia is a long-night plant. Flowers massively come out when sap starts to flow again at a period when nights are still long enough for inducing flowering. Then, when nights have become too short, forsythia ceases to be able to produce flowers and produces leaves instead. Close to the winter solstice, when nights are longest, the conditions are particularly suitable for inducing flowering. However, the plants then being quiescent have the resources for producing only a small number of flowers. Then, when sap has begun to flow again, the plants can produce an explosion of flowers.

Box 2.2 ***Asperugo procumbens***
Have you ever heard of that plant? Probably not. It is a species of plant that almost nobody would recognize if coming over it. I have met it however. I have even grown it for some time.

I was Assistant at the laboratory of plant physiology of the Faculty of Sciences of Paris, then housed at the Sorbonne. The Director of the laboratory, Professor Pierre Chouard, was a learned botanist, whose pleasure was to gather plants wherever he was. Once he took some of his collaborators with him for a studious trip in the Pyrenees. I was among them. We passed with him unforgettable moments. Not only he knew by name all the plants that we met, including the less frequent, but he told us about their evolution, their ecology, the specific features of their physiology and a quantity of other details. One morning, he led us to a rock, the appearance of which seemed to present nothing special, and yet he asked us to examine it carefully. He finally told us that what grew under the rock was the famous *Asperugo procumbens*, a rare plant with absolute photoperiodic requirements: it is induced to flowering only after exposure to sufficiently short nights. This is the reason why it blooms only during the period around the summer solstice (approximately May to August). We took photographs, picked up a few seeds, but did not cut off any sample of the plant in order not to deteriorate the site.

Most of the protagonists of that trip are now dead. Those still alive are too old to go wandering again in mountain, and, no doubt, they have long forgotten where the plant was. It is likely that *Asperugo* is still under its rock, as one of these pirate treasures of which nobody today knows the cache.

may then occur when night length is slightly above or slightly below the flowering threshold, depending on the local climatic conditions. In the first case, the seedlings will immediately produce a few flowers and stop growing, thus remaining dwarf plants all their life. In the second case, instead of immediately flowering, the plants will have an abundant vegetative growth and will finally flower in autumn when the night length will be above the threshold. The two types of plants thus obtained will look totally different from each other. Only expert botanists can then understand that they both belong to the same species!

By contrast with the long-night plants, the short-night plants (Box 2.2) are induced to flowering when the night length is below a given threshold.

2.2.2 *Plant Sensitivity to Gravity*

On earth, the aerial parts of plants grow upwards and the roots downwards (which is termed a "gravitropism"). If you lay a young plant horizontally (Fig. 2.2a), stem and root will bend in such a way as to grow again upwards and downwards,

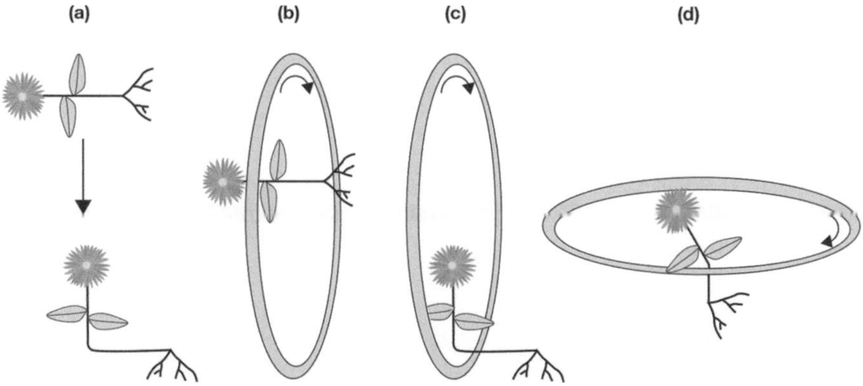

Fig. 2.2 The effect of gravity and of a centrifuge force (Knight's wheel). (**a**) A plant that is arranged horizontally (under such conditions that the root does not dehydrate) redirects its shoots upwards and its roots downwards. (**b**) A plant installed horizontally and perpendicularly to the plan of a vertical wheel which rotates very slowly (so that the centrifuge force remains negligible) retains its orientation. The reason is that during the rotation of the wheel, gravity being successively in all directions, its resulting effect is nil. (**c**) If the wheel turns rapidly, the effect of gravity remains negligible, but the plant redirects its growth, the roots growing in the direction of the centrifuge force and the aerial parts in the opposite direction. (**d**) If a plant is installed vertically and perpendicularly to the plan of a horizontal wheel rotating at medium speed, it redirects its growth according to the resulting effect of gravity and the centrifuge force

respectively. Studying the consequences on plant growth, of either getting rid of the effect of gravity or creating an artificial gravity by application of a centrifuge force, has clearly demonstrated that such behaviour of plants was a response to gravitation.

A conclusive (but costly) way to get rid of the effect of gravity consists of experimenting under weightless conditions within an artificial satellite. In that case, whatever the initial orientation given to a young plant, the plant keeps it and continues to grow without bending.

Knight's wheel is a more economical way to proceed. Consider first a vertical wheel turning more or less rapidly and arrange a young plant at the periphery of the wheel with any possible initial orientation: gravity acts on the plant in all directions in a lap of the wheel, thus nullifying its resulting effect. If the wheel turns slowly enough for creating no significant centrifuge force (Fig. 2.2b), again the plant keeps its initial orientation and continues to grow without bending. When the wheel turns more rapidly, gravity remains inefficient while the centrifuge force, by creating an artificial sort of gravity, ceases to have a negligible effect. The plant then bends in such a way as to reorient the growing root in the direction of the centrifuge force and the growing stem in the opposite direction (Fig. 2.2c). Lastly, consider a young plant arranged in any possible position on a horizontal wheel. In that case, gravity is no longer ineffective. If the wheel turns rapidly enough for inducing a significant centrifuge force, the growing plant will bend in such a way as to orient the root growth in the direction of the resultant effect of gravity and centrifuge force and the shoot growth in the opposite direction (Fig. 2.2d).

Fig. 2.3 Female *Bryonia dioica* plant (Photo: Marie-Claire Verdus)

2.2.3 Local and Distant Responses to Stimulation

Bryony *(Bryonia dioica)* is a dioecious[1] cucurbitaceous plant, the long, thin stem of which clings to bushes and gets entangled in them (Fig. 2.3). Bryony is a common plant, usually found in hedges in our regions. It bears nice-looking but toxic red fruits with the size of redcurrants. The stem elongates by forming successive internodes carrying alternate leaves. The terminal, newly formed internode elongates rapidly, while a new internode is formed, then elongates rapidly and so on during the entire growth of the plant.

Control bryony plants are obtained by maintaining them, from their germination, sheltered from wind, rain, touching or any other sort of stimulation. If the terminal, growing internode of such a plant is gently rubbed during its elongation, this internode, when fully elongated, will be shorter and thicker than the corresponding internode of the control plants; moreover, the equipment of the stimulated internode in peroxidase enzymes will be modified compared with the corresponding internode of control plants (Boyer et al. 1979). By contrast, neither the growth of the next internodes nor the peroxidase equipment of both the younger and older internodes is modified. This is a typical example of a local response (i.e. modifying the metabolism and growth of only the stimulated internode).

[1] A dioecious plant is a non-hermaphroditic plant: there are distinct, male or female, bryony plants.

As soon as 1935, Roger Gautheret (who was also famous for his work on plant tissue culture) has studied the case of a response occurring at a distance from the stimulated area (Gautheret 1935). The experiment consisted in arranging a young wheat plant in such a way as to maintain the root wet (in order to prevent plant drought) and in observing the growth of the root tip with a microscope equipped with a micrometre. More precisely, Gautheret estimated the rate of root growth by measuring the time passed for the root tip to cover the distance between two neighbouring graduations of the micrometre. At a given instant, he administered a pinching or cutting stimulus 1 cm above the root tip. After a short interval of time, the rate of root growth changed abruptly. This is thus a case in which the growth of the root tip is modified as a consequence of administering a stimulus at a distance of 1 cm.

Since then, other examples of distant responses have been observed. For instance, inflicting excisions, abrasions or punctures to young pea plants rapidly caused a modification of the systems of protein biosynthesis[2] at a distance of up to 15 cm from the wounded area (Davies and Schuster 1981).

2.2.4 *Defence Reactions of Plants*

The defence reactions of plants to wounding by herbivores (insects, molluscs or mammals) and to infestation by pests (fungi, bacteria or viruses) are many. A few particularly dramatic examples are given below.

Plants have invented the chemical war long before us. When attacked by herbivores, various plants respond by the biosynthesis and accumulation in their foliage of molecules (e.g. alkaloids, tannins) that are toxic for these predators and discourage them to continue. It is noteworthy that this remarkable adaptation is beneficial for both the attacking herbivore and the attacked plant. The herbivore can feed on the plant, but it cannot browse until total destruction of the plant (since this plant becomes progressively more and more toxic and thus more and more dissuasive). Reciprocally, the plant can save on the synthesis of defence molecules as long as no dangerous herbivore has begun to chew it.

Plants have also invented the biological war. In the simplest case, trees (e.g. *Macaranga*, *Acacia*) host ants and feed them with nectar or other nutritive substances that they produce. Reciprocally the ants protect the trees from herbivores by biting them (Heil et al. 1997; Young et al. 1997). In other cases, when they are attacked by herbivores, some plants emit pheromones[3] or other volatile substances that attract predators (or parasites) of the attacking herbivore (Dicke and van Loon 2000; Allmann and Baldwin 2010). The signalling pathways are progressively deciphered (Mewis et al. 2005).

---

[2]The modification consisted in the arrangement of ribosomes into polysomes (for detail see Sect. A2.6).

[3]Pheromones are volatile substances that are liable to change the behaviour of the animals and/or plants that are sensitive to them.

Fig. 2.4 *Calligonum*. This photo was taken on the great erg in Sahara. The *Calligonum* has adapted itself to the dry conditions of the desert by possessing a double root system. The main root plunges deep into the sand to find water, and a network of surface roots catches the water provided by the condensation that occurs early in the morning

Especially in desert areas, some plants would even be able to warn neighbouring plants that herbivores might come to them. Deserts are far from being abiotic areas. Some plants, such as *Calligonum*, succeed to grow on the great erg in Sahara (Fig. 2.4). Animals are present too. They are often extremely modest, but herds of antelopes and small groups of dromedaries also live there (Fig. 2.5) and feed on rare herbs and bushes. When the herbivores begin to browse one of these bushes, it has been reported that not only this bush starts synthesizing toxic substances but also starts emitting pheromones (or other volatile substances such as ethylene) that trigger the synthesis of toxic substances by the neighbouring bushes even before they are attacked. Many reports of this kind exist in the literature. Some have been disputed. However, laboratory experiments, carried out under conditions that seem reliable (Karban et al. 2000), support the idea that plants can really communicate with other plants via the emission of volatile substances (Karban and Baldwin 1997).

Plants react to infestation by pathogenic microorganisms by apoptosis[4] of the cells surrounding the point(s) of contamination. This "hypersensitive" response, which is an efficacious way to prevent the propagation of a local infection to the rest of the plant, is followed by enzyme biosynthesis, cell wall hardening and other metabolic modifications that eventually lead to a systemic resistance to the invasive pest (Pontier et al. 1998).

[4] Programmed cell death.

Fig. 2.5 A small herd of dromedaries in the desert

Fig. 2.6 The Venus fly trap (*Dionaea muscipula*), a carnivorous plant (Photo: Jeanine Rens). The three hairs that trigger the closure of a Dionaea trap are not discernible in this figure

2.2.5 Plant Movements in Response to Stimuli

When a plant responds to a stimulus by a movement, again this response may be local (capture movement of the Venus flytrap) or distant (folding movement of the mimosa).

The Venus flytrap (*Dionaea muscipula*) is a relatively rare, carnivorous plant, the leaves of which are arranged across a circle with a radius of approximately 5–10 cm (Fig. 2.6). It lives in acidic bogs in Eastern USA. Such media are usually poor in

Box 2.3 Diabolical Venus Flytrap

The traps of the *Dionaea* are impressive, but the way how they act (capture and digestion of a prey) is even more dramatic. There are two levels of discrimination enabling the plant to capture only preys that are worth being captured.

Within each trap there are three small hairs that are sensitive to touching. The trap will not close if a hair is touched only once. It will close if the same or another hair is also touched within a few dozens of seconds. It is unlikely that this double touch occurs when some inert object (e.g. small gravel or tiny sprig) falls on an open trap by accident. By contrast there is a high probability that an insect that alights on a trap will move and touch one or several trap hairs in its movements, thus triggering the trap closure.

When a trap has started closing, it does not continue until complete closure but stops midway for a short time and then continues to close. This means that a fly (or any other living being with a comparable size) will remain caught, whereas midges have time to escape. When a trap has completely closed on a fly, the process of digestion takes approximately 10 days. When the prey (or preys) was small enough to escape at mid-closure, the trap reopens in the course of 1 day.

What is the advantage for the Venus flytrap to possess such a sophisticated mechanism?

First, capturing and digesting an insect is a long process that is costly in energy. Second the traps can function not more than approximately three times. It is thus an advantage for the plant not to close its traps on a nonedible or too insignificant prey.

nitrate. To compensate the shortage of its nitrogen nutrition, the Venus flytrap captures and digests small insects. At the end of the leaves, the traps are made of two lobes the edges of which are equipped with long, rigid filaments. When a fly alights on a trap, the Venus flytrap closes the lobes, thus imprisoning the fly (Box 2.3). Enzymes are secreted by the internal face of the lobes, which hydrolyse the proteins of the prey into amino acids and oligopeptides (short sequences of amino acids) that are absorbed by the plant. This is thus a case in which a movement (closure of the lobes) and a metabolic process (secretion of digesting enzymes) constitute a local response to the touching stimulus caused by the insect.

Mimosa (Fig. 2.7) is a common plant in tropical areas, which is also easily grown under temperate climate. The slender stem bears alternate leaves. These leaves are made of a primary petiole that divides into four secondary petioles, each of which bears a series of leaflets attached to the secondary petioles by tertiary petioles. Touching or any other weak stimulus causes the folding of the stimulated leaflet and of a few neighbouring leaflets. Progressively increasing the intensity of the stimulus causes the folding of all the leaflets of the secondary petiole, then the folding of the secondary petiole, the folding of the four secondary petioles, the folding of the pri-

Fig. 2.7 The sensitive plant (*Mimosa pudica*) (Photo: Jeanine Rens)

mary petiole and the folding of the entire leaf. Finally the most injurious stimuli can cause the folding of several leaves. After a while, the leaves come back to their initial position. In each of these cases, the propagation of the movement of the stimulated area to more distant areas is easily visible. This comes to actually see the propagation within the plant of the signal induced by the stimulus.

The mechanisms driving movements are rather complex. In mimosa, humps (termed "pulvini") are present at the base of the primary, secondary and tertiary petioles. Pulvini are made of a heap of cells that can modify their swelling by changing (via chemical reactions or transport of appropriate substrates) the concentration of solutes in their internal medium (especially their vacuoles). These modifications of swelling are responsible for a deformation of the pulvini that drives the movement of the corresponding petioles.

2.3 How Do Plants Perceive Stimuli?

Plants do not possess sensory organs similar to our ears, eyes, nose, etc. How do they perceive the stimuli to which they are sensitive?

2.3.1 Perception of Light Signals

Plant sensitivity to light depends on the existence of chromoproteins (proteins associated with a nonprotein pigment), one of which, termed "phytochrome", has been intensively studied. Phytochrome plays a major role in plant sensitivity to the relative length of night and day. To better understand the way how this molecule operates, let us begin with a brief reminder of the nature of light.

The most familiar light is daylight, which appears to us homogeneous and "white". This means that this light does not exhibit any particular colour. However, those who have seen a rainbow or observed daylight through a prism know that daylight can be split into a sequence of coloured lights classically termed "purple-indigo-blue-green-yellow-orange-red". This sequence however is a rough approximation. No boundary exists between well-defined colours. For instance, there is continuity from a quasi-green blue to a quasi-blue green, then progressively to a quasi-yellow green, and so on. The white light is in fact the combination of an infinite number of coloured lights, each being extremely close to the preceding and following ones. Each of this infinity of coloured lights is said to be "monochromatic" (i.e. corresponding to a single colour).

Now, let us come back to phytochrome. Under laboratory conditions, plants can be illuminated with either of two different monochromatic reds, the first being close to orange and the other close to infrared. Depending on which of these two monochromatic lights is used, the plant phytochrome evolves between two isomeric[5] forms that we shall term *phytochrome 1* (illumination with the quasi orange red) and *phytochrome 2* (illumination with the quasi infrared red). When the phytochrome of a *Xanthium pennsylvanicum* plant is in the form *phytochrome 2*, the plant is induced to flowering.

Under natural condition of illumination (i.e. under daylight), it is more complicated. The two forms of phytochrome coexist, and both are continually degraded and resynthesized at different rates. When night length is at least equal to 9 h, the relative content of *phytochrome 2* reaches a value permitting the *Xanthium pennsylvanicum* plant to be induced to flowering.

2.3.2 *Perception of Gravitation*

The gravity stimulus, and/or comparable stimuli caused by some other forces (e.g. the centrifuge force), acts on the whole plant. Hence it is not possible to focus these stimuli on any specific organ of the plant in order to determine the site of perception.

It seems however that the perceiving areas, for roots and shoots, are located close to the tip of these organs. The reason is that removing the terminal part of both organs abolishes the plant sensitivity to gravity. Moreover special cells (termed "statocytes") are visible in the root tips and slightly below the shoot tip. These cells contain little, starch-rich bodies (termed "statoliths") that settle on the cell mem-

[5] The atomic composition of two isomeric molecules is the same, the atoms being however arranged differently. For instance, the molecules CH_3CH_2CHO and CH_3COCH_3, which are both made of three carbon atoms, six hydrogen atoms and one oxygen atom, are isomeric: the first is an aldehyde and the second a ketone.

brane in the direction of gravity alone or of the combination of gravity with a centrifuge force.

In the classical interpretation, (1) the interaction of the statoliths with the part of the cell membrane on which they have settled activates membrane proteins; (2) this activation of proteins causes an ubiquitous plant hormone (termed "auxin") to be asymmetrically distributed, which eventually (3) orients the growth of the shoots upwards and that of the roots downwards. Some observations support that interpretation. In particular, the stems of statocyte-defective mutants have lost the ability to hold themselves erect and merely creep on the ground (Roberts 1984; Gaiser and Lomax 1991). However, other observations do not fit perfectly well with the classical interpretation, which might mean that this interpretation is still uncomplete.

2.3.3 *Perception of Diverse Stimuli*

We have seen above that volatile substances, such as ethylene, were involved in plant-to-plant signalling. This is just one effect of ethylene among many others. Ethylene acts on a lot of physiological processes (germination, tissue senescence, fruit ripening, leaf fall, etc.). It seems, at least in some cases, that ethylene interacts with various plant hormones, including auxin, and thus reorients the plant metabolism. Generally speaking, the perception by plants of a signalling substance begins with the specific binding of this substance to a protein that is termed the "receptor" of this substance. Receptors are located on the plasma membrane of target cells of the signalling substance under consideration. The complex "receptor/signalling substance" engages the plant in a series of processes leading to its ultimate response.

For many other stimuli that are perceived by plants, the initial steps of perception are still incompletely understood. However, it is likely that the perception of a mechanical stimulus passes via modifications of membrane tensions, that of a thermal shock via changes in the rate of functioning of metabolic processes and that of a drought stress via the alteration of the cellular medium viscosity (Victor Norris, personal communication). We shall see below (Sects. 3.6.1 and 4.2.3) that laying droplets of a solution of diverse substances on plant leaves has an effect comparable to puncturing these leaves, which might mean that these substances play a part in the perception of wounding.

Rain, wind and touch induced the expression of calmodulin and calmodulin-related genes in *Arabidopsis* (Braam and Davis 1990).

2.4 After Stimulus Perception, the Calcium Wave

Between stimulus perception (Sect. 2.3) and the ultimate response of a plant, a series of reactions and processes take place that are still not totally unravelled.

However, one of the stages in this series has been thoroughly studied (Box 2.4). Immediately after the perception of diverse stimuli (including touching, wind, cold

Box 2.4 Evidencing the Calcium Wave by Use of Genetically Modified Plants

The way how a British team (Knight et al. 1991) has studied the transient invasion of the cytosol by calcium as a consequence of stimulus perception (termed the "calcium wave") is elegant and worth being related. It is based on introducing a luminescence gene into the plant genome.

Everybody knows the glow-worm that can be found in the grass at nightfall, though it is not as frequently encountered nowadays as it was formerly because of light pollution of the atmosphere and the massive use of pesticides in agriculture. The glow-worm is not at all a worm but the female of a coleopteran insect (*Lampyris noctiluca*). The female has no wings and thus cannot fly in search for a sexual partner, but it is able to emit flashes of light that attract males (which can fly).

This property to emit light, termed "bioluminescence", is not restricted to the glow-worm. A number of bioluminescent species exist. They belong to families (microorganisms, insects, molluscs, fishes, etc.) often far from one another in the systematic classification of living beings. For instance, *Aequorea victoria* is a jellyfish in which a protein (termed aequorin) transforms, in the presence of calcium, a part of its metabolic energy into light.

The British team succeeded to graft the aequorin gene in the genome of a few plants (including *Arabidopsis thaliana*, the favorite experimental material of plant biologists). The aequorin protein was thus present in the cell cytosol, which enabled it to emit light when it contained a sufficiently large concentration of calcium. In the absence of any stimulus, the calcium concentration of the cytosol is too low to trigger the emission of light, but when the plants perceive a stimulus, a flash of light reveals the invasion of the cytosol by calcium. Moreover, the properties of this light flash (duration, intensity, kinetics) permit to estimate the corresponding properties of the calcium wave.

shock, drought stress, etc.), a sudden but transient increase of the concentration of free calcium,[6] termed a "calcium wave" (Trewavas 1999), occurs in the cytosol of the cells involved in the perception of the stimulus. More precisely, in the absence of any stimulus, the concentration of free calcium in the cell cytosol is very low (on the order of 1 μg per litre, i.e. 1 millionth of a gram per litre), while it can be a hundred thousand (100,000) times higher in the nutrient solution[7] of the plant and in

[6] Calcium may be present under three different forms in cell media. The "free calcium" diffuses without restriction, the "fixed calcium" is chemically bound to proteins or other molecules and the "condensed calcium" is mobile along unidirectional, negatively charged molecular assemblies but cannot move away from these assemblies even at infinite dilution (for more detail, see Appendix A3).

[7] Under natural conditions, plants usually grow on the soil; however, under laboratory conditions, it is often more convenient to grow them in an appropriate solution of mineral salts that is termed "nutrient solution" or "nutrient medium".

intracellular pools such as vacuoles and mitochondria. On stimulus perception, the cytosol connects with these calcium-rich pools, thus causing its invasion by calcium. Shortly afterwards, the cytosol disconnects from the calcium-rich pools, and "molecular pumps" convey calcium back to exterior and to the internal pools until the cytosol is brought back to its initial, calcium-poor state.

It has been observed that the amplitude, kinetics and duration of the transient elevation of the cytosolic calcium were dependent on the type of stimulus perceived by the plant (touching, cold shock, etc.). This is interpreted as meaning that the transient elevation of cytosolic calcium orients the plant towards an appropriate response to the perceived stimulus (Dolmetsch et al. 1997; McAinsh and Hetherington 1998; Knight et al. 1998). More precisely, this calcium wave, via calcium-dependent enzyme proteins, is assumed to trigger a series of processes leading to epigenetic (Sect. A2.7) modifications of gene locking/unlocking. DNA methylation/demethylation reactions have indeed been observed to take place in plants subjected to stimuli and stresses (citations in Thellier and Lüttge 2013). As a consequence of this locking/unlocking of genes, the production of novel proteins, the disappearance of previously existing proteins, the opening of membrane ionic channels, etc. can occur and lead towards the ultimate response.

2.5 From the Stimulated to the Reactive Area

When the response occurs in a tissue different from that perceiving the stimulus, it is obvious that information has to be transferred from the stimulated to the reactive area.

In the experiments by Roger Gautheret (Sect. 2.2.3), a method for estimating the rate of this transfer of information might consist of dividing the distance between the stimulated and the reactive areas by the time lapse between stimulation and response. In doing so, there is however the risk to underestimate the rate of information transfer. The reason is that the time lapse between stimulation and response depends not only on the time lapse for information transfer but also on the time lapse for intermediary stages between information arrival at the root tip and the growth response of the root to occur. Therefore the method under consideration does not really measure the rate of information transfer. It rather gives a lower limit for this rate. When all the calculations were done, it was found that the lower limit of the rate of information transfer was on the order of one to a few tenths of a mm per second in Gautheret's experiments.

The mechanism for information transfer in plants has been the subject of intensive studies and disputes. The main possible mechanisms are (1) a cargo of signal molecules in the sap flow, (2) the propagation of a hydraulic pressure wave and (3) the propagation of an electrical depolarisation wave. Depending on the plant species and on the type of stimulus under study, cases may occur in which several of these mechanisms operate simultaneously. It may be too that it is now one and now another of these mechanisms that is involved.

It is usually admitted that an electrical process is involved in the closure of the *Dionaea* traps and in the folding of the mimosa leaves. In both cases, the rate of information transfer has been estimated to be on the order of one to a few mm per second, which is appreciably more rapid than the above-calculated value for the experiments by Roger Gautheret. This might mean that the lapse of time for processing intermediary stages between the arrival of the signal in the reactive area and the plant response is shorter when this response is a movement than when it is a metabolism or growth modification. If this is actually the case, then the response to movement permits a better estimate of the rate of information transfer in plants than the metabolic or growth responses do.

2.6 About a Few Practical Applications

The knowledge concerning plant sensitivity to various types of stimuli has already led to practical applications.

The oldest applications deal with plant sensitivity to mechanical stimuli. It has been orally reported that long ago, in India, the cotton farmers went through a strange ceremony during which they whipped the growing cotton plants. Although the author has not found any written document supporting this assertion, this does not prove that the anecdote is not truthful. It may be that the whipping induced the ramification of the cotton plants, thus increasing the cotton yield.

Closer to us, in the 1970s, several groups have adjusted devices for avoiding an excessive elongation of the stem of plants grown in greenhouses (Hammer et al. 1974; Beyl and Mitchell 1977; Wheeler and Salisbury 1979). For instance, it was commercially important that the stem of chrysanthemums (and other posies) was neither two short nor too long. This was achieved by setting the pots containing the growing plants on vibrating stages and controlling the vibration frequency and the duration of the treatment so as to obtain chrysanthemums with an optimal size. Similarly, by subjecting tomato plants to regular agitation or water sprinkling, it was possible to reduce the stem elongation thus improving the plant resistance and vigour. All this was obviously achieved without any genetic manipulation or utilization of chemical products.

Since "Maryland Mammoth" tobaccos are long-night plants that do not flower when night length is below 10–11 h, (Allard and Gabner 1941; Allard 1947), these plants can be maintained in a vegetative state by imposing periods of daylight larger than 14 h or by interrupting long nights with brief periods of illumination. This increases the production of leaves, which is commercially advantageous.

Plant defense substances, such as alkaloids, are commonly used for medical application.

Chapter 3
Discovery of the Existence of Memory in *Bidens* Seedlings

According to Shepherd (2005), the Indian biophysicist Jagadis Chandra Bose (1858–1937) concluded from a series of experiments and bioelectrical measurements that plants were capable of remembering and learning. This was not well received in his time and was rapidly forgotten.

The history of the rediscovery of plant memory is full of interest but rather intricate. If readers have too much difficulty with the present chapter, they may simply skip it and go directly to Chap. 4 in which simpler experimental systems are described.

3.1 Correlations Between Organs

The nowadays approach of plant memory began with the study of the correlations existing between the different organs of a plant. Decisive contributions were due to Marie-Odile Desbiez, prematurely departed at the beginning of the year 1995. At the University of Clermont (France), in the CNRS[1]-associated laboratory managed by Professor Paul Champagnat, she was the head of a small team. They were studying the complicated network of interactions causing that the growth and metabolism of each plant organ are influenced by what happens in other organs. Marie-Odile's favourite plant material was *Bidens pilosus* (recently feminized as *Bidens pilosa*). The reason for the name of that plant is that the seeds possess two long spikes or "teeth" ["*dents*" in French]. The total number of *Bidens* plants that Marie-Odile treated and observed one by one in the course of her short life is on the order of 1.5 million!

[1] Centre National de la Recherche Scientifique [National Centre for Scientific Research].

M. Thellier, *Plant Responses to Environmental Stimuli*,
DOI 10.1007/978-94-024-1047-1_3

3.2 Dominant and Dominated Buds

Frequently, a hierarchy becomes established with dominant and dominated individuals in a group of higher animals (a simple flock of cows, for instance). This is very well-known. We are going to see that a somewhat similar behaviour is observed between the buds of a plant. However, let us begin with recalling an observation familiar to any gardener.

Consider a twig at the moment when it starts growing again in spring. At the tip of the twig, the terminal bud (or "apex") grows actively, thus elongating the twig and producing new leaves. At the same time, the other buds present on the twig remain quiescent. This is termed "apical dominance", which means that the terminal (or "apical") bud "dominates" the other buds (i.e. prevents them to be active). The way how the apex exerts its dominance has been disputed. Does it consume most of the resources available, thus depriving the underlying buds of nutrients? Does it emit signal substances (e.g. hormones) inhibiting the growth of the other buds? Can these two mechanisms co-exist? Whatever it is, removing the apex (by sectioning the extremity of the twig) causes the underlying buds (or, at least, some of them) to immediately become active and create sub-ramifications carrying leaves.

3.3 Specification of the Dominance Between Cotyledonary Buds

Marie-Odile Desbiez began by verifying that the behaviour of young *Bidens* plants was fairly analogous to that of the twig. At the age of one to a few weeks, the structure of a *Bidens* seedling (Fig. 3.1) is comparable to that of the young radish plant depicted in Fig. 1.1. Each seedling is made of an axis (root, hypocotyl and apex), two opposite cotyledons and two cotyledonary buds (each at a cotyledon axil). The cotyledonary buds, dominated by the growing apex, normally remain small and inactive. However, as soon as the apex has been removed (plantlet "decapitation"), they immediately start to grow. Marie-Odile Desbiez then went through a series of three experiments.

In the first experiment (Fig. 3.2a), the *Bidens* plants were grown *under optimal conditions of light and mineral nutrition*. When the plants where decapitated using a razor blade at the onset of daylight (it will be seen below why these precisions are important), both cotyledonary buds of each plant started to grow. They grew at approximately the same rate, which means that neither of them dominated the other one.

In the second experiment (Fig. 3.2b), the same operations were carried out, but with plants grown *under conditions of low light and poor mineral nutrition*. When

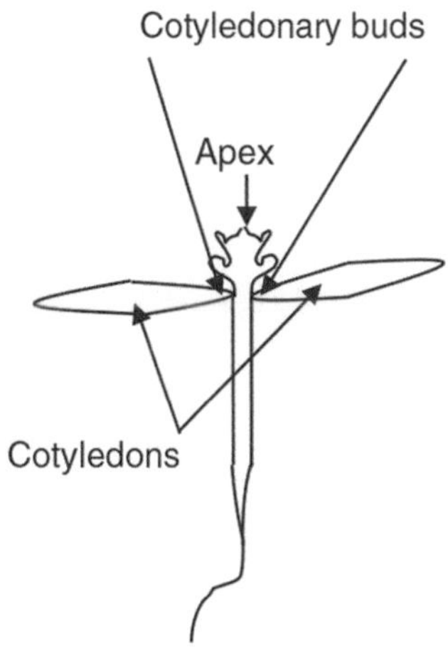

Fig. 3.1 Scheme of a *Bidens* plantlet at the age of 2–3 weeks. This is the initial state of the plants in the experiments described in Sects. 3.3 and 3.5

the plants were decapitated (again using a razor blade at the onset of daylight), both cotyledonary buds of each plant started to grow, but one of the two buds started growing before the other (i.e. dominated the other). Therefore, under the unfavourable conditions used, as soon as plant decapitation had removed apical dominance, a new relation of dominance was established, this time between the cotyledonary buds themselves. However, it was impossible to predict which of the two cotyledonary buds of each plant would become dominant. In other terms, each of the two cotyledonary buds had equal chance to become dominant.

In the third experiment (Fig. 3.2c), the plants were treated as in the second one, except that an *asymmetrical stimulus* (pricking one of the two cotyledons with a blunt glass needle) was administered *at the moment when plant decapitation was carried out* (once more in the morning and using a razor blade). Again one of the two cotyledonary buds of each plant started to grow before the other, but the bud at the axil of the pricked cotyledon became dominant in appreciably less than 50% cases (i.e. the bud at the axil of the non-pricked cotyledon was dominant in more than 50% cases).

On the whole, under the experimental conditions as indicated, administering an asymmetrical pricking stimulus to *Bidens* seedlings causes the dominance relation between the two cotyledonary buds to change from a situation in which it is undetermined to a situation in which it is specified. In brief, the administration of the asymmetrical stimulus results in the specification of the dominance between the cotyledonary buds. Moreover the "degree of the specification effect" is measured quantitatively by the percentage of plants (50–100%) in which the cotyledonary bud at the axil of the non-pricked cotyledon is dominant.

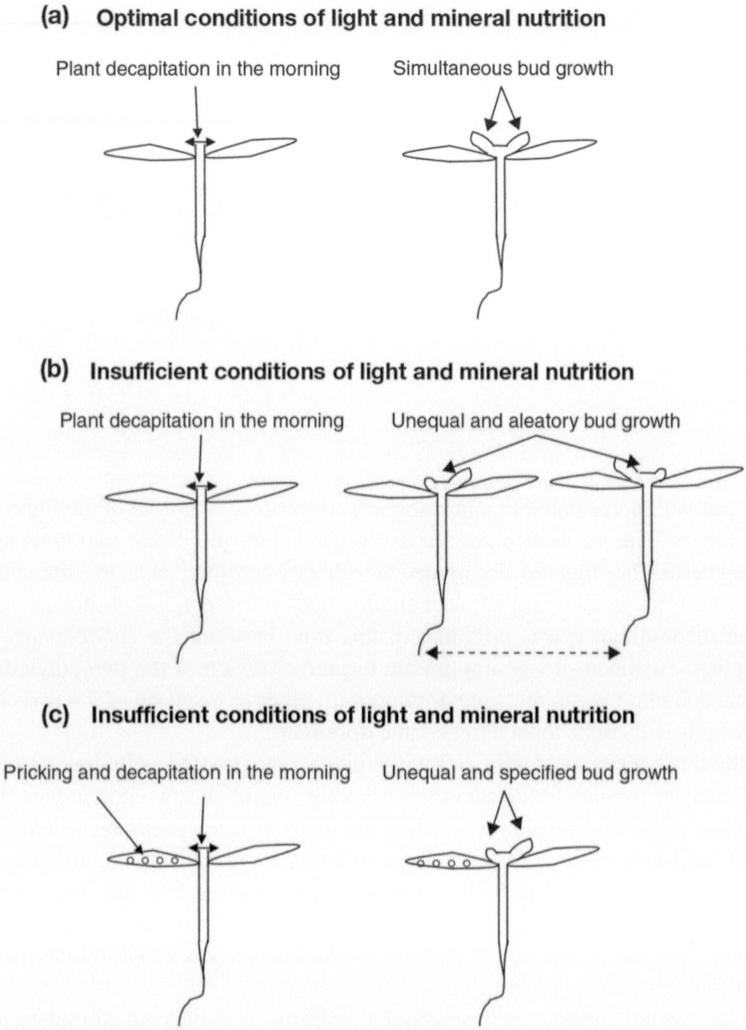

Fig. 3.2 Relative growth of the cotyledonary buds of *Bidens* seedlings after plant decapitation. (**a**) Under optimal conditions, the two cotyledonary buds grow at approximately the same rate, i.e. neither of them becomes dominant. (**b**) Under nonoptimal conditions, one bud starts growing before the other one (i.e. it becomes dominant), but the two buds have equal chances to become dominant (non-specified dominance). (**c**) Again under nonoptimal conditions, if the plants are asymmetrically stimulated (e.g. by pricking only one cotyledon) and if this stimulation is carried out simultaneously with plant decapitation, then the cotyledonary bud at the axil of the non-stimulated cotyledon becomes dominant in most cases (specified dominance)

3.4 Rate of Information Transfer

The value of the rate of information transfer from the pricked cotyledon to the cotyledonary buds, under experimental conditions similar to those in the preceding section, can be estimated as follows.

Bidens seedlings were again decapitated simultaneously with being subjected to an asymmetrical pricking stimulus. Then both cotyledons were removed[2] shortly after this double treatment. When the time lapse between pricking and cotyledon removal was on the order of a few dozens of seconds, the dominance between the two cotyledonary buds was undetermined. By contrast, the bud at the axil of the non-pricked cotyledon was statistically dominant (i.e. the bud at the axil of the pricked cotyledon was dominated) when the time lapse was up to 2–3 min. A possible interpretation is that (1) information specifying the dominance induced by the asymmetrical pricking treatment had still not reached the cotyledonary buds when the time lapse was too short, while (2) 2–3 min were needed to cover the distance from the pricked site on a cotyledon to the cotyledonary buds. The distance being approximately 1–3 cm, it is easily calculated that the rate of information transfer is on the order of one to a few tenths of a mm per second. This is consistent with the value that was determined from Gautheret's experiments (Sects. 2.2.3 and 2.5).

Furthermore, one can detect and study the propagation of electrical signals using extracellular electrodes.[3] When one of the cotyledons of a *Bidens* seedling was subjected to a pricking stimulus, two different electrical signals (termed "action potential" and "slow wave") were propagated from the stimulated to the reactive area. The rate of propagation of the action potential was appreciably higher than that estimated for information transfer (Sect. 2.5). By contrast, the rate of propagation of the slow wave was on the same order of magnitude as that estimated for information transfer. Moreover, the amplitude of the slow wave was higher at the level of the bud at the axil of the pricked cotyledon than it was at the level of the opposite bud. Without being fully convincing, all that suggests that the slow wave might be the vehicle for information transfer in the young *Bidens* plant (Frachisse et al. 1985).

3.5 Storage/Recall of Dominance-Specification Information

The discovery of the existence of a memory involving information storage and recall was the result of a large number of experiments carried out and interpreted in a collaboration involving the author and Marie-Odile Desbiez (Thellier et al. 1981,

---

[2] It might seem that removing the cotyledons is an extremely strong stimulus, capable of deeply disturbing the experiment. In fact, this is not the case. What is important when a plant is wounded is the number of injured cells rather than the surface of removed tissue. The clear ablation of the cotyledons with a razor blade thus is a traumatic stimulus much weaker than the pricking of a cotyledon with a blunt needle.

[3] In animals the propagation of electric signals occurs via nerves. In plants, it occurs via phloem cells. The ions involved are not the same in animals and plants.

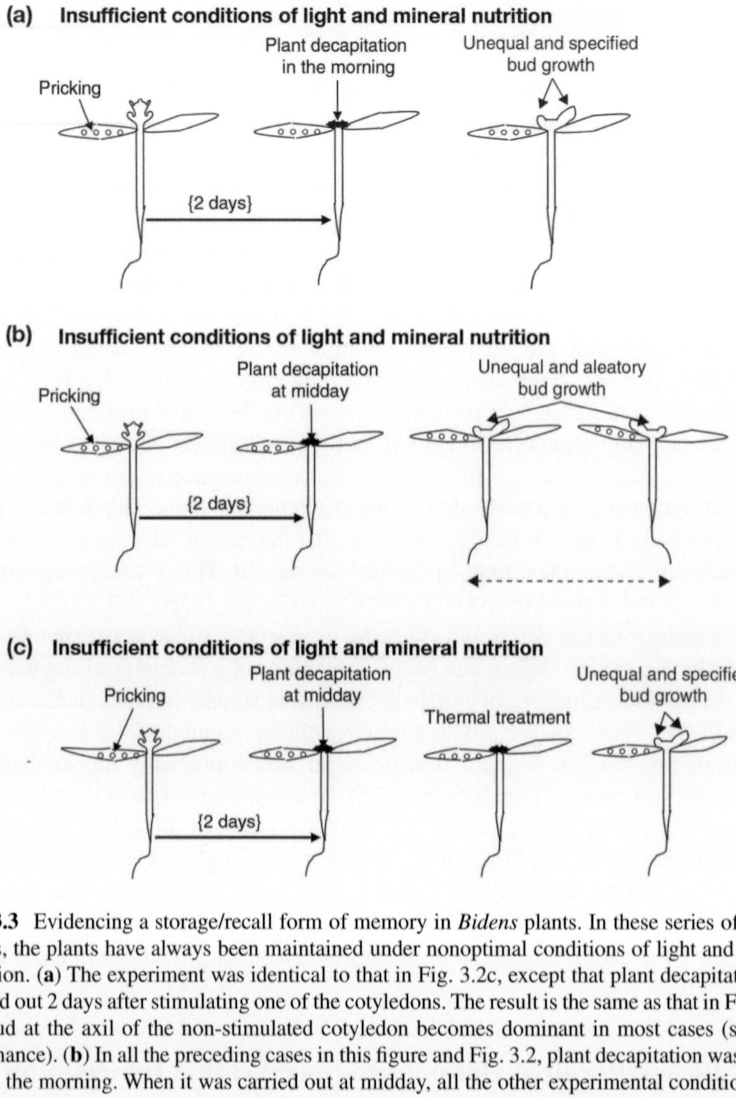

Fig. 3.3 Evidencing a storage/recall form of memory in *Bidens* plants. In these series of experiments, the plants have always been maintained under nonoptimal conditions of light and mineral nutrition. (**a**) The experiment was identical to that in Fig. 3.2c, except that plant decapitation was carried out 2 days after stimulating one of the cotyledons. The result is the same as that in Fig. 3.2c: the bud at the axil of the non-stimulated cotyledon becomes dominant in most cases (specified dominance). (**b**) In all the preceding cases in this figure and Fig. 3.2, plant decapitation was carried out in the morning. When it was carried out at midday, all the other experimental conditions as in (**a**), bud dominance was non-specified, but (**c**) when subjecting the plants to a non-asymmetrical thermal treatment immediately after plant decapitation, bud dominance was afresh specified

1982; Desbiez et al. 1984, 1991). In Sect. 3.3 we have seen (Fig. 3.2) that, when *Bidens* seedlings were *simultaneously* decapitated and subjected to asymmetrical pricking stimulation, this caused the *specification of the dominance* between the cotyledonary buds. The question then arises: what will happen when plant decapitation is retarded with regard to the application of the asymmetrical stimulus? Only one series of three experiments (Fig. 3.3) will be described here. These experiments were carried out using *Bidens* seedlings in the same initial state as depicted in Fig. 3.1 of Sect. 3.3.

In the first experiment (Fig. 3.3a), the experimental protocol was similar to that described in Fig. 3.2c except that the pricking stimulation and plant decapitation were 2 days apart from each other instead of being carried out simultaneously. The result however was the same. When plant decapitation occurred, the cotyledonary buds immediately started to grow. The growth rate of the two buds was not the same, and the dominant bud was most often that at the axil of the non-pricked cotyledon. Hence, at the moment when apex removal permitted the cotyledonary buds of the plants whose decapitation was delayed to start growing, these plants “remembered” that they had undergone an asymmetrical stimulation 2 days earlier and they responded to this stimulation by specifying the dominance between the cotyledonary buds (Thellier et al. 1981). Hence, information relative to dominance specification between cotyledonary buds (1) has been stored within the plants and (2) has remained latent during the 2 days between the asymmetrical stimulation of the plant and the start of bud growth.

The second experiment was intriguing (Fig. 3.3b). When plant decapitation was performed (using a razor blade) no longer at the onset of daylight but at midday, all the other experimental conditions being identical to those in the preceding experiment, the two cotyledonary buds had equal chances to become dominant. Bud dominance was thus undetermined when the plants were razor-decapitated at midday instead of being specified when the plants were razor-decapitated in the morning. In more detail (by putting together published and unpublished data[4]), when the plants were decapitated with a razor blade in the morning, at midday and in the evening, bud dominance was specified, undetermined and specified, respectively. Moreover, when the terminal bud was torn out with pliers instead of being neatly removed with a razor blade, the reverse occurred (bud dominance then being undetermined, specified and undetermined). This was unexpected: why did the conditions for removing the terminal bud (which was a symmetrical operation) have effect on the memorization of an asymmetrical stimulus? Could some of the different manners to carry out plant decapitation have the effect of rubbing off the memorization of the asymmetrical, pricking stimulus?

The third experiment (Fig. 3.3c) clarified the problem. The experimental conditions were identical to those depicted in Fig. 3.3b except that the plants were subjected to a thermal treatment (rapid cooling and slow return to room temperature) shortly after being razor-decapitated. The bud at the axil of the non-pricked cotyledon was then dominant, i.e. bud dominance was again specified. This means (1) that information induced by the stimulus was always *stored* within the plants (i.e. it was never erased) but that (2) depending on the experimental conditions (e.g. mode of plant decapitation, possible addition of a thermal treatment), the plants were either able or unable to *recall* stored information and let it take effect by specifying or not the dominance between the cotyledonary buds. Information stored as a consequence of the perception of an asymmetrical pricking stimulus was maintained without any appreciable erosion up to 2 weeks (a time lapse after which the plants have grown so much that they are no longer valid for this type of experiments).

[4]The unpublished data are personal communications by Marie-Odile Desbiez shortly before she died.

This is what we have termed a storage/recall form of plant memory. To some extent, this resembles the storage/evocation form of animal memory. Both in plants and animals, memory cannot be restricted to information storage. A recall function is requisite; otherwise stored information would indefinitely remain latent and serve to nothing. As an example, in our everyday life, we have memorized (i.e. stored) the pathway from where we work to where we live. Most of the time this stored piece of information remains latent somewhere in our brain (we do not even think of it). However, when we wish to go back home, we bring it back to consciousness (i.e. we "evoke" or "recall" it) to find our way.

3.6 Properties of the Storage/Recall Form of Memory

In Sect. 3.5, it has been shown that the *Bidens* plant possesses memory and that information specifying the dominance between cotyledonary buds may be stored and recalled. A few essential properties of this memory are as follows.

3.6.1 Properties of the Storage Function

Asymmetrical stimuli other than pricking have been revealed able to induce the storage of a piece of information specifying the dominance between cotyledonary buds. This was the case when administering, to a single cotyledon of each plant, traumatic (wounding) or non-traumatic (touching) abiotic[5] stimuli or else by lying, again on a single cotyledon of each plant, droplets of a solution of diverse substances. For instance, droplets of solutions of glucose, galactose, fructose or myo-inositol were active in the presence of a little potassium or calcium, while arabinose, mannitol and leucine were not, whether potassium and calcium were present or not.

In most cases, however, in subsequent experiments, we have continued to use pricking stimuli because of, both, the rapidity and easiness of their administration to plants and the quantitative character of this type of stimulus (number of pricks administered at a time). We have never found a quantitative relationship between the degree of bud-dominance specification (Sect. 3.3) and the number of pricks administered to a single cotyledon of each plant. The degree of specification randomly oscillated between 50 and 100% when the asymmetrical stimulus administered to the plants consisted of one, two, three or four pricks. Besides, once information corresponding to a pricking stimulus was stored, we have succeeded neither to erase it nor to inverse it in favour of the other bud. This remained true even when, for instance, a one-prick stimulus on the left cotyledon was followed by a four-prick stimulus on the right cotyledon. More generally when combining several pricking

[5] An abiotic stimulus is a stimulus that is not due to a living being (e.g. pricking, wind, cold shock, drought stress, etc.)

stimuli, it was always the first asymmetrical stimulus that induced information storage; the subsequent stimuli (whether symmetrical or asymmetrical) were ineffective on information storage but were possibly effective on the plant ability to recall that piece of information (see Sect. 3.6.2).

In other words, with *Bidens* seedlings, information storage, concerning bud-dominance specification as a consequence of the administration of an asymmetrical pricking stimulus, is robust (if not irreversible). Moreover, it does not depend quantitatively on the number of pricks inflicted for plant stimulation.

Using stimuli milder than pricking, the results are however different (Box 3.1). The weak specification of bud dominance that is obtained after the perception of a weak stimulus can be inversed by the subsequent administration of a pricking stimulus. Moreover the degree of bud-dominance specification is much sharper with a strong, pricking stimulus than it is with a mild stimulus.

This means that the degree of bud-dominance specification depends quantitatively on the strength of the stimulus, but that a saturation effect occurs with powerful stimuli (such as pricking treatments). This means also that the properties of the storage function that have been determined with a given type of stimulus are not necessarily the same as those determined using other types of stimulus.

Box 3.1 The Effect of Strong and Weak Stimuli

When *Bidens* seedlings underwent an asymmetrical stimulus consisting of administering to a single cotyledon of each plant, one or several pricks with a blunt glass needle, the response (specification of the dominance between cotyledonary buds) was irreversible and with an intensity independent of the number of pricks that were administered. However, even a single prick is a strong stimulus. What would occur after the administration of milder stimuli?

Marie-Odile Desbiez (personal communication) found a byway to answer the question. In her culture room, the plants in their culture vessels were arranged in such a way as having their pairs of cotyledons all parallel with one another. When facing the culture vessels, one of the cotyledons of each plant was thus on the left and the other on the right. In the culture room, there were places were the bud on the left was slightly dominant (i.e. degree of bud-dominance specification slightly above 50%) and other places where it was the bud on the right (degree of bud-dominance specification slightly below 50%). According to Marie-Odile Desbiez, this was due to small, unavoidable gradients of light and/or temperature that acted as mini-stimuli. NB: for the signification of the degree of specification, see Sect. 3.3 in the main text.

In brief, weak stimuli were responsible for a weak specification of bud dominance; then, when much stronger pricking stimuli were administered, (1) these strong stimuli imposed a sharper specification of bud dominance, and (2) they were able to inverse the specification induced by weak stimuli.

3.6.2 Properties of the Recall Function

For studying the recall of stored information, we have used not only asymmetrical stimuli (pricking a single cotyledon of each plant) but also symmetrical stimuli (simultaneously administering an equal number of pricks to both cotyledons).

As long as the lapse of time between the administration of an asymmetrical pricking stimulus and plant decapitation was not too large (for instance, 2 days), it was easy to reversibly enable/disable the plants to recall stored information. For instance, the administration of a thermal treatment, the mode of plant decapitation and the modification of the ionic state of the plants could reversibly switch the recall function "off/on" or "on/off". By contrast, with longer lapses of time (for instance, 14 days), the plants became permanently able to recall stored dominance-specification information whatever the conditions for plant decapitation and possible subsequent treatments were.

An unexpected finding was obtained by compiling the results of all the experiments in which an asymmetrical pricking stimulus was followed by another pricking stimulus after increasing intervals of time (60,000 plants studied on total). It did not matter whether the second stimulus was symmetrical or asymmetrical since, as previously indicated (Sect. 3.6.1), the second pricking treatment was ineffectual on information storage and was possibly efficient only on information recall. At first sight, the results seemed to be incoherent. In some experiments, the percentage of dominant buds at the axil of the non-pricked cotyledon was significantly above 50% (specified dominance), whereas, in other experiments, this percentage was not significantly different from 50% (undetermined dominance)! Examining the data in more detail, it was found that when the time lapse between the two stimuli was close to 1 h, 3 h, 7 h30, 12 h and larger than 14 h, then the dominance was specified, undetermined, specified, undetermined and specified, respectively. In other words, the cotyledonary-bud dominance was found to oscillate between being specified and undetermined as a function of the time lapse between the two stimuli. This was strongly suggesting that the plant ability/disability to recall stored information was somehow associated with an ultradian rhythm of the plant (Hütt et al. 2015, Lüttge and Thellier 2016). It has been indicated above (Sect. 3.5) that plant ability to recall stored information was dependent on the time of the day (morning, midday or evening) when plant decapitation was carried out, which suggests that the ability/disability for a plant to recall stored information is correlated also with a circadian rhythm.

By combining an asymmetrical pricking stimulus with one or several symmetrical stimuli, and choosing time lapses (between the asymmetrical and the other stimuli) corresponding to minimums/maximums of the oscillation described above, we have found that stored information could be repeatedly (at least twice) recalled.

3.6.3 Interaction of the Storage and Recall Functions

Clearly, it would serve to nothing to store a piece of information that will never be recalled or to try to recall information that has not been stored. Somewhat similarly to yin and yang (passivity and movement), the storage and recall of information are so complementary with each other that the existence of either of these functions would not make sense if the other wouldn't exist. Hence, it is not surprising that their combined operation has unexpected effects.

When several lots of *Bidens* seedlings were subjected to increasingly strong asymmetrical pricking stimuli (e.g. from 1 to 12 pricks administered on a single cotyledon of each plant), the result was apparently paradoxical. It was found (1) that the two cotyledonary buds had equal chances of becoming dominant (undetermined dominance) with the weakest (1 or 2 pricks) and the strongest (10–12 pricks) stimuli, while (2) the bud at the axil of the non-pricked cotyledon became dominant (specified dominance) with stimuli of intermediate strength (especially 4–6 pricks). A likely interpretation consists of assuming that some stimuli act both on information storage and on the ability/inability of the plants to recall stored information. More precisely, (1) all the asymmetrical pricking stimuli induce the storage of bud-dominance specification, (2) the plants remain in a state in which they are unable to recall stored information when they have been subjected to pricking treatments either too weak or too strong and (3) they are enabled to recall stored information only when they have been subjected to a stimulus of intermediate strength.

Whether an asymmetrical pricking stimulus was administered to young *Bidens* plants before or after having enabled the plants to recall stored information did not significantly change the percentages of dominated/dominant buds at the axils of the pricked and non-pricked cotyledons. This means that the storage and recall functions act independently from each other.

3.7 Generalizing the Concept of Plant Memory

3.7.1 Experimenting with Plants Other than Bidens

All the experiments described so far were carried out with *Bidens* plants. However, the administration of an asymmetrical stimulus induced the storage of a piece of information specifying the dominance between cotyledonary buds also in flax seedlings and in two different tomato cultivars. The ability to memorize information specifying the dominance between cotyledonary buds thus might be a fairly general property of young dicotyledonous plants.

Box 3.2 Scientific Edition
For those who are not familiar with the habits in scientific edition, let us briefly describe them. The chief Editors ask experts to evaluate the manuscripts that have been submitted to them. The experts are usually colleagues of the author(s) working in the same field. On receiving the comments of the experts, the Editors decide to publish the manuscript as it is, to reject it or to ask for amendments. The authors' names are communicated to the experts, but the experts' names are usually not communicated to the authors. This way of evaluating the qualities and drawbacks of a manuscript is not flawless and sometimes may even be unfair. However, it has revealed to be more useful than harmful, the experts' comments being honest in most cases and intending to help the authors rather than manhandling them.

3.7.2 Further Remarks about Plant Memory

As has been the case with Jagadis Chandra Bose (Shepherd 2005), our rediscovery of plant memory was initially received with reserve. We even encountered some difficulties for publishing it in international scientific journals (Box 3.2).

The initial coldness of the experts was mainly due to the nonintuitive nature of the application of the concept of memory to plants. One of them said that one might *burn his fingers* in such an affair. Another asked us about the advantage for a plant to be able to memorize information relative to the specification of the dominance between cotyledonary buds. That was a good question, which is worth being generalized as *what is the advantage for a plant to be able to memorize information*? This point will be discussed in detail in Chap. 7. Be it as it may, more and more experimental results were obtained, by us and by other teams, and the study of plant memory progressively became a research theme as accepted as any other.

The British physiologist Anthony Trewavas (2003) drew attention to the fact that at least two different forms of memory exist in plants.

- The first is the storage/recall form that we have encountered in the study of the specification of the dominance between cotyledonary buds. This form of memory implies the storage of a piece of information and the recall of that piece of information at one or several later times.
- In the second form of plant memory (as studied by several other groups), upon perception of one or several identical stimuli, plants modify (either positively or negatively) the way how they transduce another occurrence of the same stimulus. By analogy with what occurs in animals (Sect. 6.1), this may be termed a "learning" (alias "training" or "priming") form of memory. This second form of memory involves information storage. If a recall function also exists, this function is always blocked "on" because the response follows almost immediately stimulus perception.

A third form of memory (which may be termed "additive" memory) might also exist in plants. An example is given with the Venus flytrap (Sect. 2.2.5). The closure of the traps can be triggered by administering a single electrical stimulus the strength of which is equal to or above a given threshold. A series of weak electrical stimuli (the strength of each being below the threshold) also promoted the closure of the traps as soon as the sum of these weak stimuli was equal to or above the threshold value (Volkov et al. 2008).

Chapter 4
More About the Storage/Recall Form of Plant Memory

4.1 New Experimental Systems

By studying the specification of the dominance between cotyledonary buds, we have discovered the existence of a storage/recall form of memory in plants. This approach, however, is rather intricate. The cotyledonary buds are very small and buried within the plant tissues. Hence they are not easily accessible to the usual biological and biochemical methods of study. Moreover, the process under consideration (specification of the dominance between cotyledonary buds) is characterized by the percentage of dominating buds at the axil of one of the plant cotyledons; it has thus only a statistical meaning. This is indeed a complicated and disconcerting way to tackle the problem!

Fortunately (Box 4.1), two other experimental systems permitting to continue studying the storage/recall form of plant memory in a much simpler way were brought into operation. They are based on (1) the reduction of hypocotyl elongation in *Bidens* seedlings and (2) the production of epidermal meristems in the hypocotyl of flax seedlings. These two experimental systems will be described in Sects. 4.2 and 4.3 of this chapter. The first was studied by collaboration between Marie-Odile Desbiez shortly before she died and our own group in the University of Rouen. The second was studied in our group, especially by Camille Ripoll and Marie-Claire Verdus.

M. Thellier, *Plant Responses to Environmental Stimuli*,
DOI 10.1007/978-94-024-1047-1_4

Box 4.1 Discovery and Further Studies

It is not exceptional in science that a researcher, working on a given subject, fortuitously makes a fruitful discovery in a totally different domain. The word "serendipity" has been created for that. In our case it was discovering the existence of plant memory from the study of the correlations that exist between the different organs of a plant. Better known are, for instance, the histories of the discovery of vaccination by Jenner, of radioactivity by Becquerel or of penicillin by Fleming. In all cases, the discoverers had the premonition of the far-reaching consequences of their fortuitous observation. Sometimes, the way followed for the initial discovery was not well suited to the further study of the newly discovered phenomenon. However, most often, easier experimental approaches were rapidly developed.

This is more or less directly connected with another question. A few dozens of years ago, the researchers were free to choose the subject they wished to tackle. At that time, all public laboratories were financed. In many cases, such researches had no valuable consequences. However, now and then, a new way to approach a problem, an original association of ideas or an unexpected experimental result emerged and knowledge slowly but steadily progressed. Nowadays, the researchers are no longer those deciding what is worth being studied or not. Administrative authorities choose the subjects to be considered of interest and finance them exclusively. They promote preferably investigations on possible practical applications of already known principles. The national financing thus is immediately profitable.

However, imagine that such politics were followed since Palaeolithic. No doubt that we would have now exquisitely carved stones; but would we dispose also of fire, electricity, writing, decimal numeration, computers and, more generally, all the conveniences that we are accustomed to?

4.2 Reduction of Hypocotyl Elongation

4.2.1 Evidencing the Existence of Storage and Recall Functions

The experiments were carried out with *Bidens* seedlings in the days immediately following germination, i.e. with plants slightly younger than those used for studying the specification of the dominance between cotyledonary buds (Sect. 3.3). The reason is that hypocotyl elongation occurs mainly during the first days of the plant life. A simplified presentation of the experimental data appears in Fig. 4.1. Six batches of plants were numbered [a1], [a2], [b1], [b2], [c1] and [c2]. Hypocotyl elongation was measured during the 2 weeks following plant germination under different conditions of mineral nutrition and of plant stimulation (Desbiez et al. 1987).

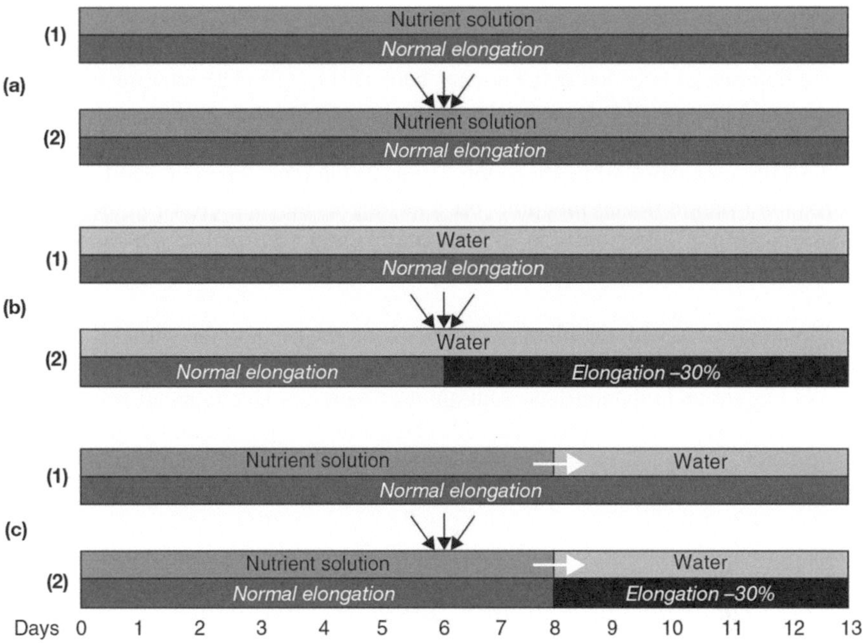

Fig. 4.1 Storage and recall of information relative to the reduction of the daily hypocotyl elongation in *Bidens* plants. The *arrows* point at the moment when the plants were stimulated

The plants of the first two batches were grown in a *conventional nutrient solution*, either in the absence of any stimulation (batch a1) or by subjecting them on the 6th day to a four-prick stimulus administered simultaneously to both cotyledons (batch a2). We did not observe any significant difference in the daily hypocotyl elongation whether the plants were stimulated or not.

The plants belonging to batches b1 and b2 were grown in distilled water (i.e. in an *extremely diluted* mineral solution), but they were treated exactly as those in batches a1 and a2, that is to say either non-stimulated (batch b1) or subjected on the 6th day to a four-prick stimulus on both cotyledons (batch b2). In that case, the daily hypocotyl elongation of the stimulated plants was approximately 30% less than that of the non-stimulated plants. Hence the four-prick stimulus, which did not seem to have any effect on the plant grown in a nutrient solution, caused a significant inhibition of hypocotyl elongation with the plants grown in water.

The plants of the last two batches (c1 and c2) were initially grown in a nutrient solution, and they were either non-stimulated (batch c1) or subjected on the 6th day to the same pricking stimulus as above (batch c2). After that, the plants of both batches were transferred into water on the 8th day. The daily hypocotyl elongation of the stimulated plants remained nonsignificantly different from that of the non-stimulated plants as long as they were both grown in the nutrient solution, but it was reduced by approximately 30% as soon as the plants were transported into water.

These results are easily interpreted by considering that two functions ("storage" and "recall") are involved in a manner similar to that in the case of the specification of the dominance between cotyledonary buds (Sect. 3.3). The interpretation is as follows: (1) the stimulus (four pricks administered simultaneously to both cotyledons) induces in all cases the storage of a piece of information about the reduction of the daily elongation of the hypocotyl, and (2) the plants remain unable to recall stored information (recall function "off") as long as they are grown in the nutrient solution, while (3) they become able to recall stored information and use it in the control of hypocotyl elongation when they are transferred into water (recall function "on").

More precisely with the plants of batch c2, on the 6th day the stimulus induced information storage relative to a reduction of hypocotyl elongation, but the plants then being grown in the nutrient solution, they were not able to recall stored information and nothing apparently occurred. Then, after being transported into water on the 8th day, the plants perfectly "remembered" that they had undergone a stimulus 2 days earlier and they responded to it by reducing the daily hypocotyl elongation.

4.2.2 *Properties of the Memorization of a Reduction of Hypocotyl Elongation*

The response is always symmetrical: the hypocotyl does not bend, whether the pricking stimulus has been symmetrical or asymmetrical. What is important is the total number of pricks. For instance, an asymmetrical stimulus of eight pricks administered to a single cotyledon has an effect similar to that of a symmetrical stimulus consisting of administering four pricks to both cotyledons. Moreover, the stored piece of information depends quantitatively on the strength of the stimulus: the reduction of the daily hypocotyl elongation, which is approximately 30% for a stimulus of eight pricks on total, falls down to approximately 15% and 7% with stimuli consisting of four and two pricks, respectively. The rate of propagation of the signal was again found to be on the order of a fraction of a mm per second (Sects. 2.5 and 3.4).

Whether the plants had been transferred to water before or after the administration of a pricking stimulus did not change the value of the daily hypocotyl elongation. This means again (Sect. 3.6.3) that the storage and recall functions operate independently from each other.

The storage as well as the recall of a piece of information dealing with the reduction of hypocotyl elongation was accompanied (Henry-Vian et al. 1995) by modifications of the biosynthesis of a few proteins characterized by their molecular weight expressed in kDa (kilodaltons). In particular, the administration of a pricking stimulus (inducing information storage) modified the biosynthesis of two proteins (37 and 41 kDa), while transferring the plants into water (enabling them to recall stored information) modified the biosynthesis of three proteins (20, 48 and 49 kDa). As expected, subjecting the plants to both the pricking stimulus and the transfer to water changed the biosynthesis of the five proteins (20, 37, 41, 48 and 49 kDa).

Moreover, this changed also the biosynthesis of two proteins (both close to 38 kDa) that are probably involved in later stages of the response. The tissue content in ethylene precursors, the rate of ethylene liberation and the activity of various peroxidases increased in the course of a few hours following the administration of a pricking stimulus to the *Bidens* seedlings (Desbiez et al. 1987).

4.2.3 *Extension to Other Plants than* Bidens

The administration of cotyledonary pricks has also induced the storage of a piece of information about the reduction of the daily hypocotyl elongation in young tomato plants grown in a classical nutrient solution. Then, transferring these plants into water has enabled them to recall stored information and let it take effect by reducing hypocotyl elongation and modifying peroxidase activities. After administering eight pricks to both cotyledons (i.e. administering a 16-prick stimulus on total), the reduction of the daily hypocotyl elongation was up to 50% (Lefèvre et al. 1993). Similar results were obtained by subjecting young plants of flax to cotyledon-pricking stimuli and transferring them into water. Moreover, lying droplets of dextran or fructose solutions laced with potassium chloride on the cotyledons of young flax plants grown in water also caused a reduction in the daily hypocotyl elongation (quotation, by Thellier et al. 2013, of a personal communication by Marie-Odile Desbiez).

So not only *Bidens* seedlings but also young plants of flax and tomato (and probably many other plants) were able to memorize information about the reduction of hypocotyl elongation. Moreover, droplets of solutions of some sugars might act as stimuli able to induce the storage of morphogenetic[1] information.

4.3 Production of Epidermal Meristems

4.3.1 *The Experimental Approach*

Flax (*Linum usitatissimum*) is a plant with nice blue flowers that is abundantly raised in Europe. The experiments relative to the storage/recall of meristem-production information were carried out using young flax plants (variety "ariane") during the month following their germination.

When flax seedlings, shortly after their germination, were subjected to, both, an abiotic stimulus and a transient shortage of calcium, we have found (Verdus et al. 1997) that epidermal cells in the hypocotyl started dividing, thus producing small meristems (Fig. 4.2). Among different sorts of abiotic stimuli, we have often used a

[1]A morphogenetic agent is an agent controlling the acquisition of the shape and structure of an organism or of a part of an organism (here, an agent controlling hypocotyl elongation).

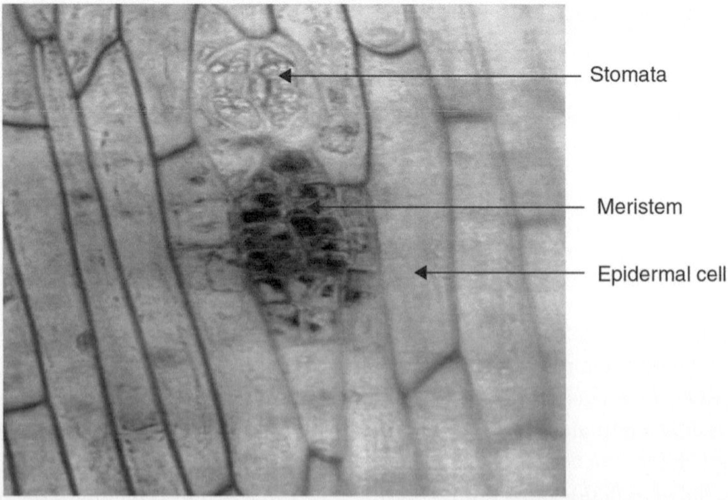

Fig. 4.2 Epidermal meristem in flax hypocotyl (Photo: Marie-Claire Verdus)

mere "manipulation stimulus" corresponding to taking the seedlings out of their germination box and putting them into a culture vessel containing a classical nutrient solution. The transient shortage of calcium was carried out by cautiously transporting the whole batch of plants into a nutrient solution, the calcium concentration of which was severely decreased or nil, for instance during 2 days, and then bringing the plants back to their normal nutrient solution.

For meristem detection, the plants were cleared by dipping in a mixture made of equal volumes of water and ethylic alcohol during 24 h, and then the meristems were spotted and counted using a low-magnification microscope.

4.3.2 Evidencing the Existence of Storage and Recall Functions

In the first experiment, flax seedlings were subjected to a manipulation stimulus *immediately followed* by a transient shortage of calcium. Every day (for instance), a dozen of plants were taken off their culture vessel, and the number of meristems produced in their hypocotyl was counted. From this counting, it was easy to calculate the mean number of meristems produced per plant at the different times when measurements were carried out. By drawing the corresponding graph (Box 4.2), it is observed that the mean number of meristems per plant increases steadily and finally reaches a plateau (Fig. 4.3). When the plants were subjected to only the manipulation stimulus or to only calcium shortage, the production of meristems was always close to zero. In a second experiment, calcium shortage, instead of being applied immediately after the stimulus, was delayed by 4 or 8 days: meristem production was delayed accordingly (Fig. 4.4).

Box 4.2 Graphical Representation of Data

A convenient way for representing data such as described in this chapter consists of expressing them graphically. Two perpendicular axes are drawn, one horizontal (termed the axis of abscissae) and the other one vertical (termed the axis of ordinates). In our case, the dates for meristem numbering will be represented on the axis of abscissae, and the mean number of meristems counted at each date of numbering will be represented on the axis of ordinates. Each numbering is represented by a point at the intersection of the vertical line passing by the date when the numbering was carried out and of the horizontal line passing by the mean number of meristem counted at this date. The advantage of a graph, compared with the mere Table of experimental results, is that it permits to visualize in a glimpse the evolution of the process under study.

All this may seem complicated. However, any parent with a baby has made the graph of the evolution of the baby's weight as a function of the baby's age. The above is not more complicated than that.

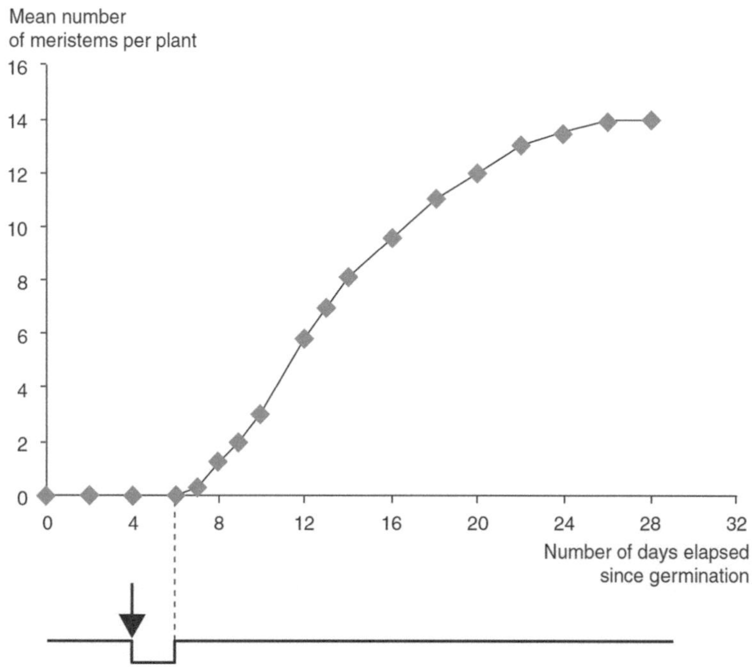

Fig. 4.3 Production of epidermal meristems in the hypocotyl of young flax plants (From Verdus et al. 1997). The plants were subjected to a manipulation stimulus at the age of 4 days (day 0 = beginning of germination), and then they were deprived of calcium from the 4th to the 6th day. The mean number of meristems produced per plant increased up to a plateau after about 3 weeks. Symbols: the moment when the plants were stimulated is indicated by the vertical *arrow* and the period of calcium deprivation by the downward indentation

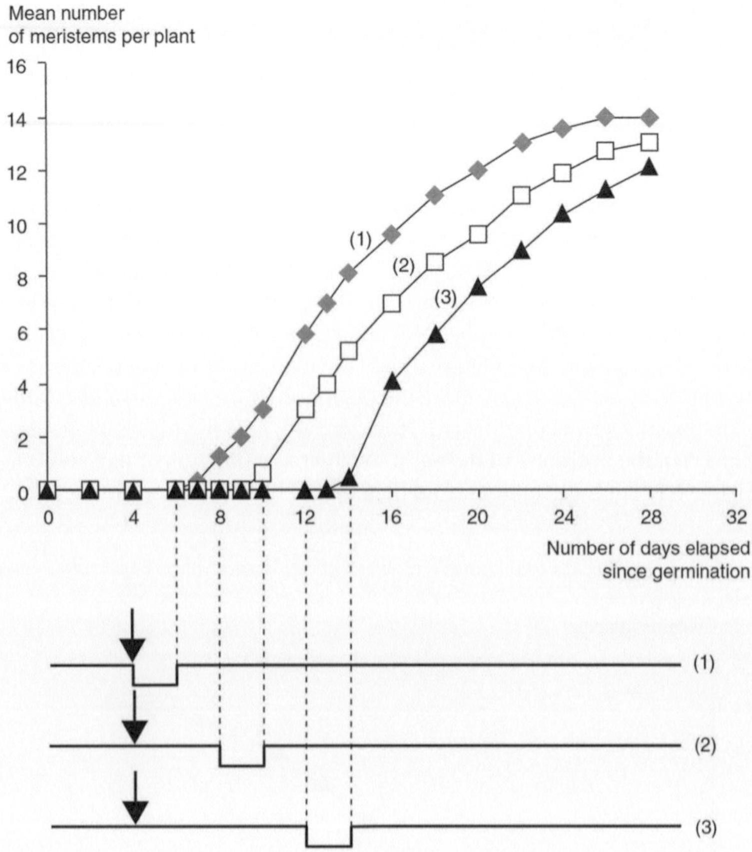

Fig. 4.4 Effect of a delay in the transient shortage of calcium on meristem production (From Verdus et al. 1997). Three batches of plants were subjected to a manipulation stimulus at the age of 4 days. Batch n° 1 was the same as shown in Fig. 4.3. In batches n° 2 and 3, the calcium shortage was delayed by 4 and 8 days, respectively. Calcium deprivation thus took place from the 4th to the 6th day in batch n° 1, from the 8th to the 10th day in batch n° 2 and from the 12th to the 14th day in batch n° 3. The production of meristems was retarded by approximately 4 and 8 days in the plants of the second and third batches (curves n° 2 and 3, respectively) compared with what it was in the first batch (curve n° 1). Same symbols as in Fig. 4.3

Again, all this is interpreted by considering that two functions ("storage" and "recall") are involved in a manner similar to that in the cases of the "specification of the dominance between cotyledonary buds" and of the "reduction of hypocotyl elongation" (Sects. 3.3 and 4.2.1). The stimulus induces the storage of "meristem-production" information. Stored information remains latent until transient calcium shortage enables the plant to recall it and let it take effect by the production of epidermal meristems in the hypocotyl (Verdus et al. 1996). It is noteworthy that the three curves in Fig. 4.4 tend toward the same plateau (on the order of 14 meristems per plant). This means that delaying calcium shortage of 4 or 8 days causes no significant loss of information.

Whether the transient calcium shortage was applied before or after plant stimulation did not change the production of meristems. Once more (Sects. 3.6.3 and 4.2.2) this means that the storage and recall functions act independently from each other.

4.3.3 Properties of the Storage Function

Many different types of stimulation other than the manipulation stimulus have the potential to induce storage of information relative to the production of epidermal meristems in flax seedlings. For instance, our group has shown that various abiotic stimuli (such as drought, wind, cold and even irradiation by electromagnetic radiation at 0.90 GHz or 105 GHz) had a similar effect.

That a plant can be sensitive to radiations at 0.90 GHz, such as used in mobile phones (Tafforeau et al. 2002b, 2004), has been confirmed by other groups (Roux et al. 2006; Vian et al. 2006 and others cited in Grémiaux et al. 2016). It is difficult to understand where this plant sensitivity to high-frequency electromagnetic radiation comes from. Using mobile phones is far too recent for plants to have genetically adapted to the radiations that they emit. Could a natural process have an indirect effect enabling plants to perceive these radiations? In fact, we have no satisfactory explanation for the origin of this unexpected sensitivity. We can only note that it exists and add that nothing, in the experiments carried out so far, demonstrates or invalidates the idea that exposing plants to these radiations may have a deleterious effect on them.

When a series of drought shocks was administered to plants a few days after a manipulation stimulus, the production of meristems was enhanced compared with that of the plants having undergone only the manipulation stimulus (Fig. 4.5). This means that the piece of information stored after the first stimulus did not remain unchanged but was modulated on perception of the subsequent stimuli. Said differently, this also means that the response to a stimulus depends on the history of the plant.

When pharmacological agents blocking the availability or the cell absorption of calcium were present in the nutrient solution of flax seedlings, they prevented the plants from storing information concerning meristem production (Verdus et al. 2007). This result is consistent with the finding (Sect. 2.4) that stimulus perception is associated with an increase of the concentration of calcium in the cell cytosol (which, at least for a part, involves the absorption of calcium by the cells). The above result also suggests that the storage of information concerning meristem production somehow depends on this calcium wave in the cytosol.

Adding the same pharmacological agents in the nutrient solution of the plants after time lapses of a few minutes after a stimulus has revealed the existence of a period of vulnerability during which information storage is prevented. Beyond this period of vulnerability, information storage takes place whether the pharmacological agents are present or not. The period of vulnerability was found to be below 2 min with mechanical stimuli and above 5 min with the other abiotic stimuli that we have used (cold shock, for instance).

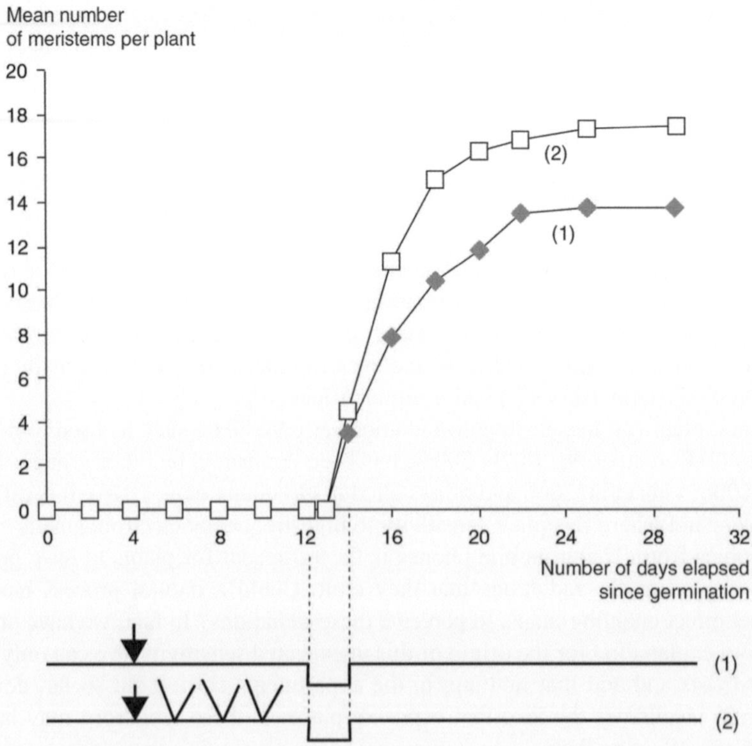

Fig. 4.5 Administration of several stimuli (From Verdus et al. 1997). Two batches of plants were subjected to calcium shortage from the 12th to the 14th day. Previously, the plants of batch n° 1 had undergone a manipulation stimulus on the 4th day. The plants of batch n° 2 had undergone not only a manipulation stimulus on the 4th day but also three drought stimuli on the 6th, 8th and 10th days. The three drought stimuli are represented by spikes in V; other symbols as in Fig. 4.3

4.3.4 Properties of the Recall Function

The normal calcium content of the nutrient solution of the plants was approximately 90 mg/L. It has been indicated above that a transient shortage of calcium enabled flax seedlings to recall and express stored information concerning meristem production (Sect. 4.3.2). More precisely, the calcium content of the nutrient solution used for an effectual calcium-shortage treatment has to be below 9 mg/L. Moreover, the duration of this calcium-shortage treatment had to be at least a half day long (but less than 3 days in order not to damage the plants). Furthermore (Verdus et al. 2012), when a transient calcium shortage has enabled plants to recall stored information (recall function "on"), a subsequent, transient excess of calcium prevented them to recall stored information (recall function "off"). The calcium-excess treatment consisted of, for instance, transferring the flax seedlings into a nutrient solution containing 270 mg calcium per litre. In brief, it is possible to reversibly enable/disable flax seedlings to recall stored information by sufficiently decreasing/increasing,

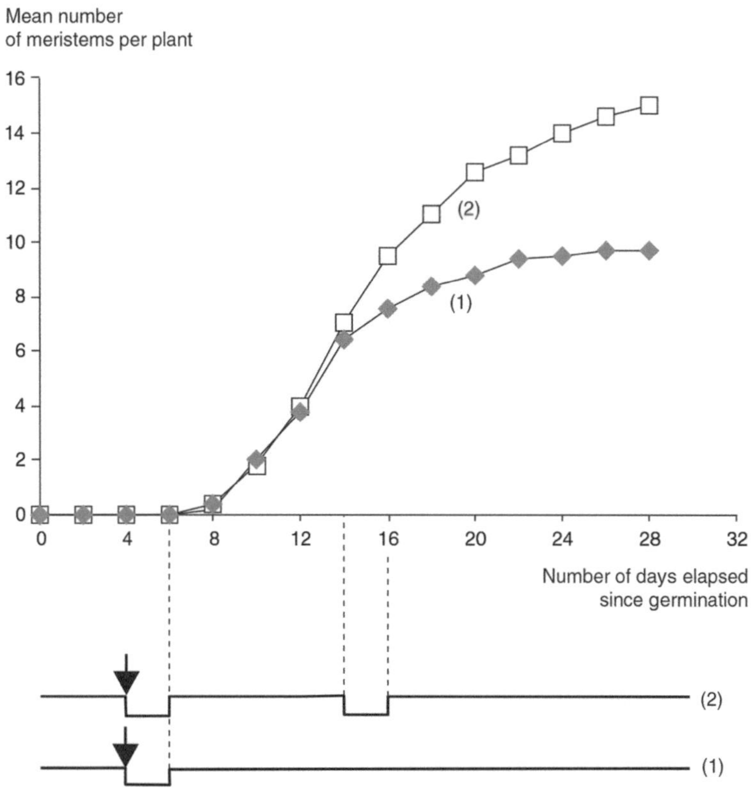

Fig. 4.6 Effect of a second deprivation of calcium on the production of meristems by previously stimulated flax seedlings (From Verdus et al. 1997). Two batches of plants were subjected to a manipulation stimulus at the age of 4 days. In batch n° 1, the plants suffered a single shortage of calcium from the 4th to the 6th day. In batch n° 2, the plants also suffered a shortage of calcium from the 4th to the 6th day, and then they suffered a second shortage of calcium from the 14th to the 16th day. The production of meristems, which normally reached its plateau in batch n° 1, resumed in plants from batch n° 2 immediately after the second deprivation of calcium. Symbols as in Fig. 4.3

respectively, the concentration of calcium in their nutrient solution. Two other experiments yielded an interesting result.

In the first experiment (Fig. 4.6), flax seedlings were subjected to a manipulation stimulus immediately followed by a transient shortage of calcium. As usual they produced epidermal meristems in their hypocotyl (curve n° 1). When meristem production had practically reached its plateau value, subjecting the plants to another depletion of calcium caused a new round of meristem production to occur (curve n° 2).

In the second experiment (Fig. 4.7), again when flax seedlings were subjected to a manipulation stimulus immediately followed by a transient shortage of calcium, they produced epidermal meristems in their hypocotyl (curve n° 1). When seedlings, initially treated exactly as the preceding ones (stimulation immediately followed by

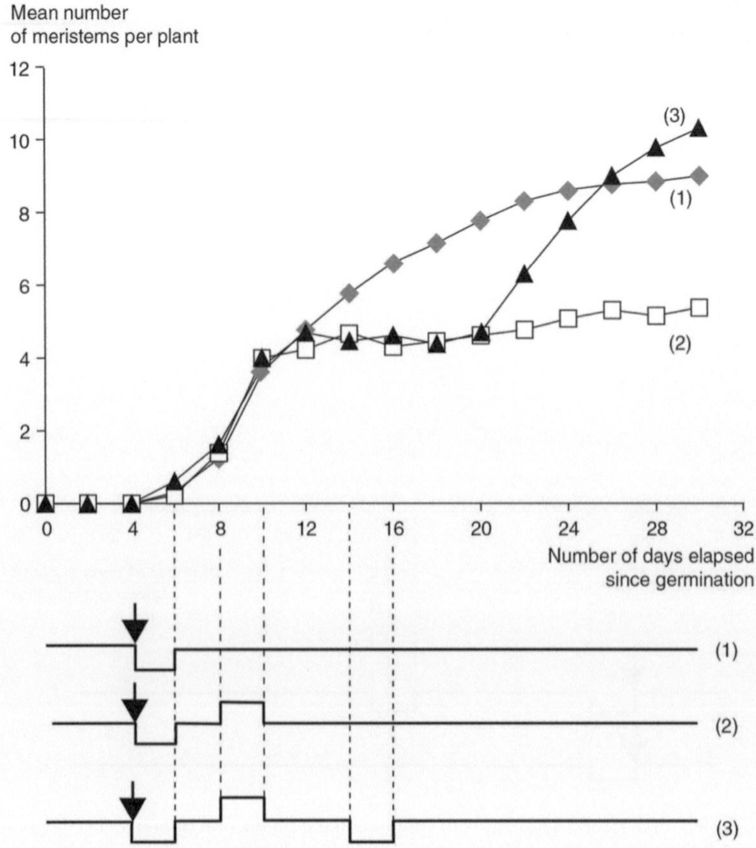

Fig. 4.7 Effect of combining calcium deprivation and excess on the production of meristems by previously stimulated flax seedlings (From Verdus et al. 1997). Three batches of plants were subjected to a manipulation stimulus at the age of 4 days. In batch n° 1, the plants suffered calcium shortage from the 4th to the 6th day. In batch n° 2, they also suffered calcium shortage from the 4th to the 6th day, and then they underwent calcium excess from the 8th to the 10th day. In batch n° 3, they suffered calcium deprivation and excess as in batch n° 2, and then they suffered a second deprivation of calcium from the 14th to the 16th day. The production of meristems, which normally reached its plateau in plants from batch n° 1, was almost completely stopped after the period of calcium excess in batch n° 2, but resumed after the second deprivation of calcium in batch n° 3. The period of calcium excess is indicated by the upward indentation; other symbols as in Fig. 4.3

calcium shortage), were subjected to calcium excess, meristem production was interrupted (curve n° 2). When seedlings, initially treated exactly as the latter (stimulation, calcium shortage, calcium excess), were subjected to another calcium shortage, a new round of meristem production occurred (curve n° 3).

These two experiments agree to show that, again (Sect. 3.6.2), stored information can be recalled repeatedly (twice in our present case).

4.3.5 *Memory and Rhythms*

The extent of meristem production depends on the period of the year when the experiments are carried out. This is the reason why the plateau of the curves n° 1 (curves corresponding to experiments carried out at different dates) has not the same value in Figs. 4.4, 4.5, 4.6 and 4.7. In spring (i.e. in the period April–June), a non-negligible number of meristems is produced in the non-stimulated and/or non-calcium-deprived controls; however, the number of meristems produced in the stimulated and calcium-deprived plants is always at least ten times larger than it is in the controls. This again (Sect. 3.6.2) suggests that the memorization process might be associated with a plant rhythm (an annual rhythm in this case). Since non-stimulated controls produce some meristems in spring, the recall function rather than the storage function is likely to be connected with this rhythm.

4.3.6 *Memory and the Proteome*

The protein equipment of flax seedlings subjected to stimulation and/or calcium shortage was studied (Tafforeau et al. 2002a, 2006) using two-dimensional electrophoresis.[2] Stimulating young *Bidens* plants caused new spots to appear and other spots to be displaced (for instance, as a consequence of the phosphorylation of the corresponding proteins). Some of these protein modifications were transient and occurring early, whereas others were delayed and longer-lasting (thus probably corresponding to precocious or later stages of the response, respectively). Some of these protein modifications were specific to the stimulus administered (e.g. touching, cold shock or irradiation by electromagnetic radiation), while others were common to several different stimuli. Calcium shortage also caused a few spots to be displaced or to appear de novo.

[2] Two-dimensional electrophoresis (also termed "2D electrophoresis") is used to separate and possibly identify the proteins contained in a mixture such as a protein extract deriving from a plant tissue. This technique consists of promoting the migration (from top to bottom and left to right on an appropriate support) of the proteins contained in a drop of the mixture. The protein migration is driven by the migration of an appropriate solvent in the presence, or not, of an electric field. The various proteins then migrate more or less rapidly, depending on their molecular mass and their electric charge, and are finally revealed as a distribution of spots all over the support surface. When comparing two extracts, the presence of certain spots in the electrophoresis of one extract and not of the other means that the corresponding proteins are synthesized in the plant tissue from which the first extract derives while they are not synthesized in the plant tissue from which the second extract derives. If a given spot is not exactly at the same place in the two electrophoreses, this means that the corresponding protein has undergone a posttranslational modification in one of the tissues from which the extracts originate.

Chapter 5
More About the Learning Form of Memory in Plants

As previously indicated (Sect. 3.7.2), in the "learning" form of plant memory, upon perception of one or several identical stimuli, plants modify positively or negatively the way how they transduce another occurrence of the same stimulus.

5.1 Examples of the Learning Form of Plant Memory

There are cases in which the way how previously stimulated plants change their response to a new occurrence of the same stimulus may be observed at the moment when the calcium wave takes place. In other cases, the effect of the first stimulus (or of a series of identical stimuli) is observed in the ultimate response to a new stimulus of the same nature.

5.1.1 Memory Effect at the Level of the Calcium Wave

When young tobacco plants (*Nicotiana plumbaginifolia*) were abruptly subjected to a sudden gust of wind, as usual they answered it by transiently increasing the calcium concentration in the cell cytosol. However, when the plants were subjected to a series of gusts of wind, they ceased during approximately 1 min to make this calcium response to a new gust of wind (Knight et al. 1992). Incidentally, this also means that the memorization process takes place extremely shortly after the calcium wave.

In young plants of *Arabidopsis thaliana*, the transient increase of the concentration of cytosolic calcium induced by cold shock was attenuated when this shock was preceded by prolonged or repeated cold treatments (Plieth et al. 1999).

Again with *Arabidopsis thaliana*, a hyperosmotic stress (obtained by adding a non-absorbable substance in the nutrient medium of the plants) caused as usual an

M. Thellier, *Plant Responses to Environmental Stimuli*,
DOI 10.1007/978-94-024-1047-1_5

almost immediate increase of their cytosolic calcium. This response was enhanced in plants previously subjected to hyperosmotic treatment, whereas it was lessened in plants subjected to a strongly oxidant treatment termed "oxidative stress" (Knight et al. 1998).

5.1.2 *Memory Effect at the Level of the Ultimate Response*

The first leaves of young Poaceae (formerly termed "Graminaceae") are cooped up in a sort of sheath termed their coleoptile (which is oriented vertically under natural conditions). When young Poaceae were arranged in such a way as to orient their coleoptile horizontally, they underwent a gravity stimulus (Sect. 2.2.2) that caused the tip of the coleoptile to bend upwards. When the orientation of the Poaceae was again changed in such a way as to orient the tip of the coleoptile downwards, again the coleoptile underwent a gravity stimulus. As long as the time elapsed between the two stimuli was less than 90 min, the second stimulus again caused the coleoptile to bend upwards. However, with larger lapses of time, the second stimulus became unable to counteract the effect of the first stimulus, and the coleoptile tip continued to grow downwards (Nick and Schäfer 1988).

The Indian corn (*Zea mays*) is fairly sensitive to excessive soil salinity. However, pre-exposing the plants to low-level salinity improves their tolerance to NaCl stress. Compared with non-NaCl-exposed controls (Pandolfi et al. 2016), the plants acclimated to salt exposure reacted by retaining more K^+ and less Na^+ in the roots and enhancing their ability for Na^+ sequestration in the cell vacuoles of the leaves (which results in limiting the detrimental effect of Na^+ accumulation in the shoots).

Although the microorganisms that were formerly termed "blue algae" are now considered to be photosynthetic bacteria (alias "cyanobacteria") but no longer plants, it is worth considering the case of *Anabaena variabilis* cells. When these cells were transferred from a phosphate-poor nutrient solution to solutions the phosphate concentration of which was gradually increased, they rapidly adapted to these new conditions by reducing the activity of the cell phosphate carriers (Sect. A2.1). The way how the cells adapted to the new conditions was depending on the phosphate-nutrition conditions under which these cyanobacteria were previously grown (Falkner and Falkner 2003).

5.2 Comparison of the Learning with the Storage/Recall Form of Memory

We have seen above that it was usually easy to manipulate the recall function (thus reversibly enabling/disabling plants to recall stored information) in the case of the storage/recall form of plant memory (Sects. 3.3, 3.6.2 and 4.2.1). By contrast, with the learning form of memory, if a recall function exists, this function is always blocked in a position enabling plants to recall stored information (Sect. 3.7.2). In other respects, a distinctive feature of the learning form of plant memory is that, after perception of one or several similar stimuli, plants modify the way how they react to the perception of the same type (rarely of another type) of a subsequent stimulus (Sect. 5.1).

However, we have also seen that, in the case of the storage/recall form of plant memory, there were cases in which (1) the recall function was blocked in a position always enabling (Sects. 3.6.2) or always disabling (Sect. 3.6.3) the plants to recall stored information and (2) information stored after the perception of a given stimulus could be modulated on perception of new stimuli (Sect. 4.3.3).

Therefore, cases exist in which it is not so easy to distinguish whether a learning or a storage/recall form of memory is involved. This would be consistent with the idea that these two forms of memory might not be fundamentally different from each other but would rather be two different manifestations of a same basic process. Be it as it may, in the following we are going to continue distinguishing learning and storage/recall forms of memory, however keeping in mind that the underlying mechanisms might not be totally different from each other.

Chapter 6
Plant Memory vs. Animal and Human Memory

6.1 Reminder of some Characteristics of Animal and Human Memory

A memorization/evocation form of memory exists in many animals and in humans: information (relative to objects, events, feelings, etc.) is stored (memorization) and can be recalled to consciousness (evocation) at one or several later times. For instance, the usual definition of memory given in the "Collins Cobuild English Dictionary" is that it is *the ability to retain and recall information, ideas, images and thoughts*. When old people do not find the words that they need to express themselves, they often complain of losing their memory. However, usually, it is not the memorization that is altered (the appropriate words are still present somewhere in their brain), but it is the evocation that begins not to function very well.

Besides the memorization/evocation form of animal memory, "learning" is a process by which animals modify their behaviour as a consequence of experience or of the acquisition of information about their environment. The behaviours acquired by learning can be stored and recalled. Elementary forms of learning have been studied with lower animals such as the marine mollusc *Aplysia* (Bailey and Chen 1983). A first mode of learning is termed "familiarization' (or "habituation"), a process by which animals learn to ignore harmless stimuli. For example, when a train of small water jets or touches was administered to an *Aplysia*, the animal retracted when undergoing the first stimulus but practically ceased to react after undergoing a dozen of these stimuli. By contrast, when the *Aplysia* was subjected to a series of stronger shocks (probably felt painful), it finally retracted more violently than it did after the first shock ("sensitization"). Familiarization as well as sensitization showed themselves not only in the ultimate response of the animal but also in the nervous and metabolic stages leading to this ultimate response. Learning responses have been observed also with higher animals such as mice and rats. Even with us, the humans, our everyday behaviour are results, at least for a part, of familiarization and sensitization processes.

M. Thellier, *Plant Responses to Environmental Stimuli*,
DOI 10.1007/978-94-024-1047-1_6

Short-term and long-term memories co-exist. When one looks for a telephone number in a directory, this number is forgotten as soon as the telephone number has been dialled. By contrast, everybody has the experience of old souvenirs the detail of which has remained extremely lucid. The functioning of long-term memory involves the biosynthesis of specific mRNA and proteins (Sects. A2.1 and A2.5), which does not seem to be the case for short-term memory since the latter functions in the presence of inhibitors of protein biosynthesis. In higher animals and humans, the memory of facts and events is not acquired immediately under a long-term stabilized form. A process of consolidation is requisite before it is firmly set (Dudai 2004). This process involves the transfer of signals between brain structures such as hippocampus and neocortex (Lodish et al. 2000; Lesburguères et al. 2011), which requires some time to be carried out.

6.2 Plant Memory vs. Animal and Human Memory

As has been seen above (Chap. 5), learning forms of memory exist in plants. The responses may be more or less complicated but sometimes resemble the familiarization and sensitization responses found in animals. For instance, the weakening of the cytosolic calcium wave, as observed when a tobacco plant underwent a gust of wind coming after a series of gusts of wind or when an *Arabidopsis* seedling was subjected to cold shock after a prolonged cold treatment, resembles the animal familiarization. In similar manner, the strengthening of the cytosolic calcium wave, as observed when *Arabidopsis* seedlings were subjected to hyperosmotic stress after previously undergoing a hyperosmotic treatment, resembles the animal sensitization. In those three cases, the effects observed in plants took place at a metabolic stage (the calcium wave) anterior to the ultimate response, as it has been indicated above that this may occur in animals.

In plants, the storage/recall form of memory (Chaps. 3 and 4) plays a part somewhat equivalent to that played by the animal and human memorization/evocation. Moreover, stored information may be relatively long lasting and repetitively recalled, as it is the case with animals. As in animals, (1) a possible transfer of information is associated with the propagation of electrical depolarisation waves (Sect. 3.4), (2) information storage involves protein biosynthesis (Sects. 4.2.2 and 4.3.6) and (3) long-term memorization occurs via a consolidation process, as evidenced by the existence of a period of vulnerability to suitable pharmacological agents (Sect. 4.3.3), prior to being firmly acquired.

There are therefore a surprising number of manifestations that plant and animal/human memories share in common. However, this is a matter of convergences at the functional level rather than actual similarities. The reason is that the underlying mechanisms are totally different. For instance, the propagation of electrical waves in plants (1) takes place in phloem cells (by far less specialized than are animal neurons), (2) is much slower than it is in nerves and (3) is based on the exchange of ions different from those occurring in animals. Moreover, in plants, the consolidation of long-term memory takes place in apparently banal cells without the intervention of anything comparable to the specialized brain structures (hippocampus, neocortex) that are set into play in higher animals and humans.

Last but not least, there is a fundamental difference in the potential of plant memory vs. animal and human memory. Animals, especially humans, can memorize huge information. They keep the memory of a myriad of facts, objects, events, sensations, etc. that they have individually perceived. By contrast, in plants only the first stages following stimulus perception, i.e. the features of the calcium wave (Sect. 2.4) and a few modifications of existing proteins or biosynthesis of new proteins (Sects. 4.2.2 and 4.3.6), are specific of the stimulus perceived. Then, what is finally stored is rather a sort of "instruction" governing which metabolic and/or morphogenetic response should be made to this stimulus: specification of the dominance between cotyledonary buds (Sect. 3.5), reduction of the daily hypocotyl elongation (Sect. 4.2) or production of epidermal meristems (Sect. 4.3).

In brief, plants indeed have memory, but the nature of this memory is very different from ours.

Chapter 7
What Is the Need for a Plant to Have Memory?

7.1 Straightforward Responses vs. Memorisation

A plant that perceives a stimulus may produce a straightforward response to it, that is an almost immediate response independent of the plant history. The main advantage of a straightforward response is its rapidity. This is what enables a *Dionaea* to capture a fly that has alighted on one of its traps without giving it a chance to escape. This is also what enables plants to respond to an aggression before too much damage has been caused by an assailant. More generally, the cases are many in which a straightforward response is an excellent means to react to environmental stimulation.

Nevertheless, we have seen that memory capacities existed in all the plants that have been taken under consideration. This was the case with *Arabidopsis*, *Dionaea*, Poaceae and tobacco regarding the learning form of memory (Sect. 5.1) and with *Bidens*, flax and tomato regarding the storage/recall form of memory (Chaps. 3 and 4). Thus, it is likely that many plants (if not all) have memorization capacity.

However, the question is what is the utility for a plant to have memory?

For answering this question, it will be convenient to consider separately the cases of the learning and of the storage/recall forms of plant memory, even though they might be only two variants of the same basic mechanism (Sect. 5.2).

M. Thellier, *Plant Responses to Environmental Stimuli*,
DOI 10.1007/978-94-024-1047-1_7

7.2 Potential of the Learning Form of Plant Memory

We have seen earlier (Sect. 5.1) that, in the learning form of memory, the repetition of the same stimulus could either decrease (familiarization) or increase (sensitization) the strength of the response to another occurrence of the same stimulus. By analogy with what occurs with lower animals (Sect. 6.1), it is tempting to infer that, in plants, familiarization and sensitization take place after innocuous or harmful stimulation, respectively. However, is it actually the case?

First (Sect. 5.1), moderate wind or cold shock (such as occurring in western Europe) are mild stimuli, and it has been observed that they lead to a familiarization effect, whereas an hyperosmotic stress (mimicking a drought stress), which is a much more dangerous stimulus, leads to a sensitization effect.

Second, it has been assumed (Gagliano et al. 2014) that the folding behaviour of the *Mimosa pudica* plant is initially a defensive reflex, limiting the risk for the plant to be seen and browsed, which is triggered by the soil vibrations caused by incoming heavy herbivores. However, photosynthesis is drastically reduced in the folded state. For plants growing under low-light conditions (i.e. with a naturally poor photosynthetic activity), folding might reduce photosynthesis down to a non-tolerable level. By contrast, plants growing in full light can retain a sufficient photosynthetic activity in the folded state. Accordingly, it has been observed that high-light plants indefinitely continue to fold on approach of heavy animals, whereas low-light plants rapidly learn to ignore the incoming animals, thus privileging maintaining photosynthesis at the expense of defence (Gagliano et al. 2014).

All that supports the idea that the learning form of memory has the adaptive advantages that (1) the response is as rapid as the straightforward responses are and (2) the plants can economize the cost of making a full response to innocuous stimuli (familiarization) while maintaining or increasing the intensity of the response to dangerous stimuli (sensitization).

7.3 Potential of the Storage/Recall Form of Plant Memory

7.3.1 *Potential of the Storage Function*

It has been shown earlier that, when a first stimulus was followed by other stimuli, this modified the information stored as a result of perceiving the first stimulus (Sect. 4.3.3 and Fig. 4.5). This means that the storage/recall form of plant memory has the potential to progressively tune an integrated, updated response to the entirety of the various stimuli, and their fluctuations, that the plant has perceived in the course of time. This permits a plant to adjust its metabolism and growth to the environmental, possibly awkward conditions met at the place where the seed from which it issued fell (Fig. 7.1).

Fig. 7.1 Plant growing under unusual, awkward conditions (Photo: Chantal Kerdudou). The herb Robert (*Geranium robertianum*) is a common weed found particularly in shady, cool and damp places in wastelands, paths and woods. The specimen photographed here has rooted in a tiny crack of a cemented ground at the base of two perpendicular walls. This location is very warm and dry under sunshine. When it rains the place can rapidly become much cooler and very wet, due to water pouring along the walls and accumulating on the cemented ground. The green colour of some of the plant leaves is paler than usual. This means that the plant suffers from these tough and abruptly changing climatic conditions but nevertheless succeeds to survive and bear fruit

7.3.2 *Potential of the Recall Function*

The recall function dictates stored information not to be expressed no matter when but only when appropriate stimulations, or the environmental conditions at this moment, allow it to be recalled. This gives the potential for a plant to synchronize the moment when stored information will take effect with the progress of other internal or external processes, including those related to plant rhythms (Sects. 3.6.2 and 4.3.5).

The possibility that the recall function be blocked "on" or "off" are extreme cases in which stored information is always or never recalled, whatever the other processes taking place within the plant.

Finally, the possibility that the same stored piece of information be recalled repeatedly (Sects. 3.6.2 and 4.3.4) permits the plant to adjust the expression of this piece of information to the progress of different processes and/or of processes taking place at different moments within the plant.

7.3.3 Combined Potential of the Storage and Recall Functions

In Sect. 4.2, we have seen that following the perception of an asymmetrical pricking stimulus shortly after germination, the *Bidens* plants (1) stored a piece of information relative to hypocotyl elongation, (2) recalled that information and let it take effect only when the plants were transferred into an extremely diluted medium ("pure" water) and (3) eventually produced (Sect. 4.2.2) a symmetrical response (reduction of hypocotyl elongation).

By contrast, when *Bidens* plants underwent, a few weeks later, the same asymmetrical pricking treatment as above, they (1) stored information relative to the dominance of the cotyledonary buds upon each other, (2) recalled that information and let it take effect only when the conditions for plant decapitation (and possibly other factors) were adequate and (3) eventually produced (Sect. 3.3) an asymmetrical response (bud-dominance specification).

In brief, plants have the potential to store different pieces of information, to (repetitively) recall stored pieces of information according to different experimental conditions and, consequently, to produce different responses in its different tissues and/or at different growth stages.

7.3.4 Effective Biological Role of the Storage/Recall Form of Plant Memory

Although the storage/recall form of plant memory appears to have remarkable potential, questions remain open.

7.3.4.1 Allocation of the Plant Resources

In the three experimental systems that have been studied (Chaps. 3 and 4), three different pieces of information were apparently stored ("specification of bud dominance", "reduction of hypocotyl elongation" and "production of epidermal meristems"). This seems to mean that many different pieces of information can be stored, depending on the plant species and on the type of stimulus perceived. However, it has been indicated previously (Sect. 6.2) that plants, having no nervous system, cannot store huge information. Is there a contradiction?

The work of German colleagues (Herms and Mattson 1992; Gayler et al. 2006, 2008; Gayler 2010; Matyssek et al. 2012) may help in solving this apparent contradiction. They point out that the resources (e.g. light, water, mineral nutrition) that a plant might find in nature are usually limited. It is therefore crucial for the plant to distribute these resources between its main living activities (growth, defence, flowering, etc.) so as to maximize its chances to survive and reproduce. For instance, (1) a plant must grow fast enough compared with the neighbouring plants in order that its aerial parts remain exposed to light and permit photosynthesis, but (2) it must also develop enough defence substances to prevent herbivores or pathogens to destroy its foliage, thus rendering insufficient its photosynthetic activity. In other words, the logical response of a plant is to determine what proportion of the available resources should go to ensure growth and what proportion to manufacture defence molecules.

We may thus assume, as a working hypothesis, that what is essential for a plant is to store information determining the optimal allocation of resources to its different vital activities. The pieces of stored information that we have considered so far ("specification of bud dominance", "reduction of hypocotyl elongation", "production of epidermal meristems" and probably many others) would then be mere consequences of resource allocation.

7.3.4.2 The Dilemma of Experimental Research

The adaptive interest of the ultimate responses that we have observed is sometimes questionable. That a *Bidens* plant or bryony stores information resulting in a reduction of plant elongation is not surprising. It is well known that many plants react to wind or to touch by a reduction in their size. This enables them to better withstand high winds or to cling to the surrounding bushes during their growth. However, what can be the use for a plant to store information commanding the production of epidermal meristems after undergoing a manipulation stimulus followed by transient calcium shortage? How the ability to store information specifying the dominance between cotyledonary buds can be an adaptive advantage?

This is in fact a general dilemma with the conventional experimental method (Box 7.1). Drastic conditions have to be imposed to make a "good" experimenta-

Fig. 7.2 Three blinds and an elephant (Drawing: Yannick Kerdudou)

Box 7.1 One Elephant and Three Blinds

An exhibitor of curiosities once entered a remote village with an elephant. The inhabitants crowded in to consider the pachyderm. Among them, there were three blinds who undertook to feel it since they could not see it. The first, who arrived from the rear, seized the tail and concluded that the beast had slender forms. The second, who found himself in front of the animal, grabbed the proboscis and inferred that this animal had some flexibility. The third, coming from the side, touched a paw and deduced that it was a being with a massive constitution (Fig. 7.2). The inferences of the three blinds were correct; but they were partial, had no reason to represent the main features of the animal and depended on the way in which the blinds had carried out their experimentation (by standing in front, aside or behind the elephant).

In seeking to understand the memory of plants, we are like the three blinds around their elephant. Constrained by the investigative means available to them, the blinds were unable to reach the "elephant" phenomenon in its entirety. We are, somewhat similarly, constrained by the imperatives of the experimental method. This method consists in "reducing" the system under study, i.e.in varying only one or a few of its settings and controlling all the others rigorously. To decipher the characteristics of plant memory, we have had to work on plants kept under laboratory conditions, i.e. under largely artificial conditions that are rarely (if ever) encountered in nature. It is not surprising that plants, subjected to unusual stimuli, have made responses that seem incongruous and which have left us blind to what we would have seen if we had been able to operate under natural conditions.

tion, but the knowledge that one draws from this experiment is inseparable from the experimental conditions that have been imposed!

7.3.4.3 Applicability of the Data Acquired so Far

The previous section has explained the ambiguity of the conventional experimental method as used in our present case. However, the question remains: to what extent can the properties of the storage/recall form of plant memory, which were deciphered using plants kept under laboratory conditions, remain valid for plants living in their natural habitat? Moreover, we have considered only the aerial parts of young herbaceous plants in our present study. What about the behaviour of roots, adult and senescent plants, woody species, etc.? The work of two other groups than ours may shed some light on these questions.

Box 7.2 Plant Tissue Culture
The culture of plant tissues was developed by Roger Gautheret (1935). The method uses glass tubes containing an aseptic gel made of a mixture of agar with a classical nutrient solution. In each tube a fragment of a plant organ (e.g. stem or root) is aseptically laid onto the gel surface; then the tubes are closed (again aseptically) by a plug of cotton. The cells of each implanted fragment dedifferentiate and divide, thus forming a more or less anarchic clump of embryonic-like cells termed a "callus". After an appropriate lapse of time (e.g. 1 month), each callus is aseptically taken off and divided into fragments, each of which is aseptically transported into a new tube. The cells continue to divide and form a new callus and so forth every month. The first of these steps is called a "tissue culture" and the following steps are termed "subcultures".

If the terminal, growing internode of an adult bryony (Fig. 2.3) is gently rubbed during its elongation, this internode, apart from becoming shorter and thicker than the corresponding internode of non-rubbed plants, exhibits enhanced peroxidase activities (Sect. 2.2.3). The memorization of the change of peroxidase activities was studied with tissue cultures obtained from fragments of rubbed and control internodes of bryony (Bourgeade et al. 1989). The tissue cultures were prepared (Box 7.2) according to the method by Roger Gautheret (1935). Increased peroxidase activities were found to exist in several successive subcultures derived from stimulated internodes. The information induced by the rubbing stimulus was thus stored during at least several months. Everything happened as if the recall function was always "on" meanwhile. Eventually, the peroxidase activities fell back to the levels existing in the control internodes and in the subcultures derived from them. There are still no experimental results permitting one to determine if this was due to the progressive erosion of the stored information or to the turning "off" of the recall function.

In a woody plant (*Rosa hybrida*), an appropriate stimulus inhibited the growth of buds formed after that stimulus but not the growth of buds formed before the stimulus (Grémiaux et al. 2016). A simple interpretation of this apparently paradoxical observation is that information (i.e. an instruction to change the allocation of resources so as to inhibit bud growth) was stored on perception of the stimulus but that this instruction was recalled only on initiation of a new bud.

Chapter 8
Attempt at Synthesis

8.1 Conceptual Modelling of Plant Memory

8.1.1 Statement of the Problem

Diverse substances, such as "memory metabolites" (Ueda and Nakamura 2006), molecules involved in the control of the cell cycle (Desbiez et al. 1998) or tiny bundles of mRNA termed "stress granules" (personal communication by Alain Vian, Davies et al. 2012), have been advocated to play a part in the plant memorization process. However, one can account for the major observed facts by drawing a model that interprets the memory of plants (whether of the type learning or storage/recall) as corresponding to an interaction between a few genes (Sects. A2.2 and A2.3).

A model (Box 8.1) is not a completely faithful representation of reality. It aims only at presenting the available data under a compact form that can be easily manipulated. Hence modelling plays a determining part in the understanding of how a system functions.

However, if the reasoning seems to be too abstract for some readers, they can skip the Sects. 8.1.2 and 8.1.3 and go directly to Sect. 8.2.

8.1.2 The Model

There are three possible pathways for a plant to produce its response to a stimulus, i.e. a straightforward pathway (Sect. 7.1) or the involvement of an either learning or storage/recall form of memory (Sect. 3.7.2). We have already indicated (Sect. 2.4) that the amplitude, kinetics and duration of the calcium wave are likely to orient the plant towards an appropriate response to this stimulus. We shall assume that the

M. Thellier, *Plant Responses to Environmental Stimuli*,
DOI 10.1007/978-94-024-1047-1_8

Box 8.1 Observation, Experiment and Modelling

At the beginning there is the insatiable interest of mankind for a better knowledge and understanding of the physical and biological features of the world in which they live.

Observation is the most obvious way for approaching this goal. By carefully examining things and events, one can learn much about them. Moreover, observation by the naked eye can be complemented by using more or less sophisticated techniques. For instance, in biology, electron microscopy has played a crucial part in the deciphering of the sub-μm structure of living systems.

Another way to extend the potential of observation consists in experimenting, i.e. in constraining as rigorously as possible most of the parameters of a given subject in order to get better information concerning one of them. We have already pointed out the main difficulty encountered with the experimental method, namely, that the data thus yielded are inseparable from the conditions under which they have been obtained (Box 7.1). However, it is the skill of the experimenter (e.g. by changing the constraints, crossing experiments with one another, etc.) to get to data with a broader meaning.

Now, the crude pieces of knowledge thus gathered on a given subject consist of a mixture of ill-assorted data, as obtained by researchers with differing scientific background and working with different plant species and under miscellaneous conditions. Our brain, despite its tremendous capacity, is not capable of directly handling such a jam of data. It is thus necessary to find out means for presenting these data in a form more easily accessible to our brain. This is the goal of modelling.

There exists an extreme variety of models, which can be ranked into two main types:

- A "conceptual" model provides a scheme of how the underlying processes are hieratically interconnected with one another. In a single glimpse, the reader can visualize the (still hypothetical) functional structure of the system (as an example, see Fig. 8.1).
- An actual "theoretical" model includes more or less mathematized relationships that describe quantitative interconnections between the different parts of a system and how they contribute to the dynamics of the whole system.

Models are not fixed abstract constructions. They are constantly amended by taking into consideration novel experimental results or new ideas of interpretation. As an example, the model in Fig. 8.1 already differs from the corresponding models given in recent publications (Thellier et al. 2013; Thellier and Lüttge 2013) and even in the original, French version of this book (Thellier 2015).

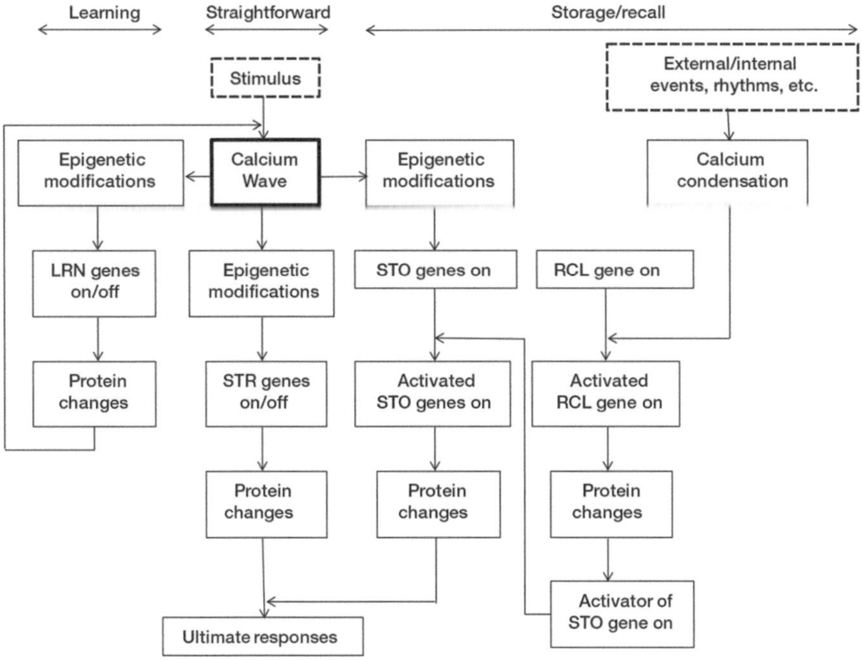

Fig. 8.1 A putative conceptual model of plant memory based on an interaction between genes. Immediately after stimulus perception, the calcium wave is an actual shunting device that orients the plant towards the most appropriate response to that stimulus. In the straightforward pathway (second column), epigenetic modifications turn on/off (i.e. unlock/lock) genes involved in the straightforward pathway (i.e. STR genes). This modifies the protein equipment of this pathway, which finally leads to the ultimate response of the plant. When learning memory is involved, the calcium wave orients the plant not only towards the same STR pathway as above (second column) but also towards an associated LRN (for learning) pathway (first column). Again epigenetic modifications turn on/off a group of genes (LRN-genes), which modifies the protein equipment in such a way as to decrease (familiarization) or enhance (sensitization) the intensity of the calcium wave. In the case of the storage/recall form of memory, once more epigenetic modifications unlock a group of genes (STO genes) that however remain inactive in the absence of a suitable activator (information storage). When the unlocked STO genes are activated, this permits the synthesis of proteins eventually, leading to the ultimate responses (third column). Besides that, various internal or external events, treatments, plant rhythms, etc. cause, perhaps via calcium condensation/decondensation processes (fifth column), the activation of an unlocked gene (RCL-gene) the product of which (fourth column) is precisely the activator requisite to activate the STO-genes. Symbols: *genes on/genes off* unlocked/locked genes, *LRN* learning, *RCL* recall, *STO* storage, *STR* straightforward

calcium wave orients the plant (1) not only towards epigenetic events (e.g. DNA methylation/demethylation) causing the locking/unlocking of genes implicated in the elaboration of a response to the perceived stimulus but also (2) towards one of the three possible pathways. The (still putative) model is as follows (Fig. 8.1).

In the straightforward pathway, the characteristics of the calcium wave induced by the perception of the stimulus orient the plant response towards the locking/unlocking of genes (termed "STR-genes", for "straightforward genes"). Then the other stages of the response (modification of the protein equipment, etc.) follow without any latency or any dependence on the previous history of the plant.

In the case of the learning form of memory (Sect. 3.7.2 and Chap. 5), after a stimulus, the progress of the intermediate stages is the same as in the straightforward case, except that the characteristics of the calcium wave orient the response not only towards the STR-genes but also towards another group of genes (termed "LRN-genes", for "learning" genes). Again, epigenetic processes lock/unlock LRN-genes, finally causing the biosynthesis of proteins catalyzing reactions that tend to decrease/increase the intensity of the calcium wave. After another occurrence of the same stimulus, the intensity of the calcium wave thus is further decreased/increased, and so on, with the consequence that the calcium wave progressively tends to vanish (familiarization) or to increase its intensity (sensitization).

When a plant follows the storage/recall pathway, the calcium wave orients the plant towards the locking/unlocking of still another group of genes (termed "STO-genes", for "storage genes"). However, the unlocked STO-genes have to be activated by an appropriate effector (termed an activator) before they can be transcribed and translated. This situation in which STO-genes have been unlocked but not activated corresponds to the storage of information. Concerning the recall function, a possibility is that the involved gene (termed "RCL-gene", for "recall gene") is always unlocked but needs to be activated for becoming functional. Diverse events (stimuli, administration of treatments, modification of environmental conditions, etc.) permit the synthesis of an activator of the unlocked RCL-gene. The stages involved remain speculative, though a calcium condensation/decondensation process (Appendix A3) and calcium-dependent enzymes might be involved. The product of this activated RCL-gene is precisely the activator of the unlocked STO-genes.

8.1.3 Adequacy of the Model

There are three main requirements for assessing the value of a model: (1) its different parts have to be consistent with one another (otherwise revising the model and/or checking the validity of the data are absolutely requisite), (2) it should contain more than what has been used to build it (emergent properties) and (3) it must be able to make predictions that can be checked experimentally.

8.1.3.1 First Requirement

No inconsistency has been found to exist either in the assumptions on which the model has been built or between the model and the observational and experimental data.

8.1.3.2 Second Requirement

The following remarks apply to the storage/recall form of plant memory.

That information storage corresponds to the proportion of locked vs. unlocked STO-genes is an emergent feature that agrees with the finding (Sect. 6.2) that the information stored is an "instruction" commanding the response to be made to a stimulus rather than the memorization of the stimulus itself.

When a second stimulus (of a similar or different nature) is added to the first, the model proposes that this produces novel locking/unlocking of STO-genes, thus resulting in creating an epigenetic state (Sect. A2.7) different from that which would have been obtained if only the first or only the second stimulus had been administered. Then a third stimulus would again alter the epigenetic state and so forth. This emergent property is consistent with (1) the experimental finding that the piece of information stored after a stimulus did not remain unchanged but was modulated on perception of new stimuli (Sect. 4.3.3) and (2) our inference that the stored information is an instruction so as to produce an integrated, constantly updated response to the entirety of the stimuli perceived in the course of time and not only to the last stimulus perceived (Sect. 7.3.1).

It has been observed experimentally that switching off/on (or inversely) the recall function (i.e. the plant ability to recall or not stored information) was sometimes easily and repeatedly reversible (Sects. 3.6.2 and 4.3.4), while in other cases, the recall function was apparently blocked "on" (Sects. 3.6.2 and 3.7.2) or "off" (Sect. 3.6.3). When considering the possibility that calcium condensation/decondensation, with its parameters ξ and ξ_c (Sect. A3), was involved in the operation of the recall function, the explanation appears as corresponding to an emergent property of the model. When the ξ value is close to that of ξ_c, even a small disturbance in the system can render the ξ value larger (or smaller) than that of ξ_c, and the recall function will then be easily turned reversibly "on" or "off". By contrast, when the ξ value is very different from that of ξ_c, it becomes difficult to render the ξ value larger (or smaller) than that of ξ_c, and the recall function will then seem to be blocked "on" or "off".

8.1.3.3 Third Requirement

An individual cell does not contain a number of copies of the same gene.

A possible interpretation for an expression such as "the proportion of locked vs. unlocked genes" may be that the cells in a tissue are not identical with each other, some being in an epigenetic state in which a given gene is locked and others in a state in which it is unlocked. Moreover, one might wonder if there are only two possibilities for a gene (to be locked or unlocked) or if cases exist in which the rates of transcription and/or translation of certain genes could vary between zero (locked gene) and a maximum (fully unlocked gene) depending on epigenetic or other processes induced by stimulus perception. The model thus predicts that one or both of these two possibilities could be found in a plant if/when appropriate techniques are available.

In the experimental part, we have always considered the effect of various stimuli on a single type of response, such as the specification of bud dominance (Sect. 3.3), the inhibition of hypocotyl elongation (Sect. 4.2) or the production of epidermal meristems (Sect. 4.3). However, increasing the intensity of one type of responses obviously increases the energy cost of that response. Therefore, in a plant having usually limited resources (Sect. 7.3.4.1), the model predicts that this plant cannot increase the intensity of one type of response without simultaneously decreasing the intensity of one or several other responses.

8.2 The Present Picture of Plant Memory

Taking into account the experiments, results and discussions that have been related so far, a fairly coherent picture of plant memory emerges. In this section, the aim is not to duplicate data already given in previous chapters but to distinguish between the points that are firmly established, those that remain speculative and mere proposals for future work. It may occur that an experimental result is well established while its interpretation is still speculative. In that case, the experimental part will be addressed in Sect. 8.2.1 and the interpretation in Sect. 8.2.2.

8.2.1 The Most Solid Knowledge

Plants are sensitive (Sect. 2.2) to many different types of stimuli (wind, touching, vibrations, electromagnetic radiation in the GHz range, injury, drought, cold shock, etc.). They react to such stimuli sometimes by a movement but most often by metabolism or growth changes. The sensors of the stimuli (Sect. 2.3) are diverse, not always well known, but in any case extremely different from our sensory organs.

Immediately after stimulus perception, a calcium wave occurs, the characteristics of which (amplitude, kinetics and duration) are specific of the perceived stimulus (Sect. 2.4). The calcium wave acts as an actual shunting device orienting the plant towards an appropriate response to the stimulus. In the three possible pathways towards the response (straightforward or involving learning or storage/recall forms of memory), the sequence of events begins with epigenetic locking/unlocking and/or activation/inactivation of genes, resulting in proteome changes. Then the proteins control the further stages towards the response.

In the learning form of memory, when a plant has undergone a series of identical stimuli, it either decreases (familiarization) or increases (sensitization) the intensity of its response to another occurrence of the same stimulus. Choosing the familiarization or the sensitization pathway depends on the nature of the stimulus. Then, the response occurs as rapidly as in the straightforward case.

The storage/recall form of plant memory is involved in the control of various processes, namely, growth (hypocotyl elongation, bud and internode growth), induction of new organs (production of epidermal meristems) and metabolic reactions (bryony peroxidases). The information stored is a sort of instruction relative to the response to be made to a stimulus, rather than the memorization of the stimulus itself (Sect. 6.2). The storage of information takes place shortly after the calcium wave, but there is a period of vulnerability (a few minutes long) before information storage finally occurs (Sect. 4.3.3). The stored information is modulated on perception of new stimuli (Sect. 4.3.3). When the ultimate response of a plant takes place at a location different from that where the initial stimulus occurred, information is stored at the level of the site of the response (Sect. 3.4). The rate of information transfer (Sects. 2.5, 3.4 and 4.2.2) from the site of stimulus perception to the site of response is on the order of 1 mm per second (perhaps a little less for metabolic or growth responses and a little more for responses in which a movement occurs). Electrical depolarization waves are involved in the transfer of information (Sect. 3.4).

Concerning the recall function, plants can often be rendered, reversibly, able/unable to recall stored information. Stored information can be repeatedly recalled (Sects. 3.6.2 and 4.3.4). The operation of the recall function is connected with ultradian, circadian and probably annual rhythms.

It does not matter if information storage occurs before or after having rendered a plant able to recall stored information (Sects. 3.6.3, 4.2.2 and 4.3.2), which means that the storage function and the recall function are independent from each other. Between information storage and the recall and exploitation of the stored information, storage periods up to a few days or even a few months have been observed (Sects. 3.6.1, 4.2.1, 4.3.2 and 7.3.4.3).

8.2.2 A Bit of Speculation

There are two possibilities concerning the learning and the storage/recall forms of plant memory (Sect. 5.2): either they are only two different aspects of the same basal mechanism or they are two completely different processes. Among the arguments supporting the first possibility, there are the facts that (1) in the learning form, the recall function (if any) is always blocked "on" (Sect. 3.7.2), while, in the storage/recall form, the recall function (though it is often easily reversible) is sometimes also blocked "on" (Sect. 3.6.2) and (2) in both cases, the response produced after perception of a stimulus depends on the history of the plant (Chap. 5 for learning and Sect. 4.3.3 for storage/recall). However, arguments also exist concerning the second possibility. In particular, in the learning case (Chap. 5), the response is always quasi immediate, and it is usually a train of identical stimuli that is effectual, whereas, in the storage/recall form (Chaps. 3 and 4), the response is always delayed and it is usually the perception of different, previous stimuli that is effectual. The question remains open to discussion. Be it as it may, it is likely that plants possess both the learning and the storage/recall forms of memory and that the functioning of these two forms is more or less connected.

What is the advantage for a plant to choose the familiarization or the sensitization pathway (Sect. 8.2.1)? By analogy with what occurs with animals (Chap. 6), it has been inferred that familiarization or sensitization enabled plants to cease to respond or to increase the intensity of their response to an innocuous or a harmful stimulation, respectively. In that interpretation, the learning form of memory appears as a remarkable mode of defence. When a plant has experienced at several instances a stimulus that has been recognized dangerous, the learning form of memory enables it to produce an intense and almost immediate response to a new occurrence of the same stimulus. By contrast, in the case of a mild stimulus, the plant can save the energy cost of producing an intense response. A few experiments support this interpretation. Moreover, would natural selection have retained the ability for plants to modulate the intensity of their response to stimuli if this would have served to nothing?

It has been reminded above (Sect. 8.2.1) that stored information is modulated on perception of new and usually different stimuli. What can be the evolutionary advantage of such a property? A reasonable interpretation (Sect. 7.3.1) is that the stored information can thus be averaged and constantly updated, which enables the plants to produce an optimized response to the myriad of stimulations (and their fluctuations) perceived in the course of time. Clearly, this would increase their chance to survive and reproduce. As Italians would say, "*se non è vero è bene trovato*".[1]

By combining the results of physiological and ecological researches (Sect. 7.3.4.1), it seems that it is via the control of the allocation of the plant resources to its main vital activities (growth, defence, resistance to stresses, etc.) that stored information induces an optimal ultimate response. It remains to be understood how exactly the control of the allocation of the resources can orient plants towards an appropriate response (such as specification of bud dominance, reduction of hypocotyl or stem elongation, production of meristems [Chaps. 3 and 4]) to the different types of stimulus. However, this is not a problem specific to plant memory but a general problem for plants with limited resources, and several German groups are actively working experimentally and theoretically on that question (Sect. 7.3.4.1).

There is a problem with the way how the recall pathway is activated. It is requisite that some sort of switch "opens the circuit". In biology, such switches often correspond to the sudden but transient increase of an ion such as Ca^{2+}, the concentration of which is normally extremely low in the cytosol. A calcium wave is the well-known process that triggers the induction of a response to a stimulus (Sect. 2.4). Could a calcium wave also trigger the start of the recall function? This is unlikely. Since there are stimuli that are effective both on the storage and the recall function, this would mean that these two functions are linked, whereas it has always been found experimentally that they are independent from each other (Chaps. 3 and 4). Apart from the calcium wave, there are not many processes by which the cell concentration of Ca^{2+} could be abruptly changed. Calcium condensation/deconden-

[1] Frequently quoted Italian proverb, meaning approximately "*If it is not true, at least it is well imagined*".

sation (Appendix A3) might be a good candidate. We thus propose the hypothesis that it is this latter process that triggers the start of the recall function. This assumption, however, remains speculative since there is no experimental support so far.

In the preceding chapters, we have implicitly considered only the case of non-heritable epigenetic characters. However, it is well-known that some epigenetic features can be maintained over several generations. This means that, when a plant has adjusted its metabolism and growth to its nearby environment, a part of these acquired properties may be transferred to its descendants. With plants the seeds of which are not scattered too far away, the descendants thus may be directly fitted to their nearby environmental conditions, without the need to go through the progressive adjustment of their metabolism and growth to these conditions (Sect. 7.3.1). Clearly, this would be an advantage.

8.2.3 Suggestions for Future Work

Much remains to be done to deepen the knowledge about plant memory and its role in plant responses to environmental stimuli.

It is requisite to carry out experiments for checking the validity of the predictions (Sect. 8.1.3) and for either rejecting the speculations (Sect. 8.2.2) or letting them become hard facts. Since most of the data acquired to date refers to the aerial parts of young, non-woody subjects studied under laboratory conditions, it remains to investigate whether one can also observe memory effects in roots, in woody species and in adult or senescent plants under conditions as close as possible to the natural ones. The stages in the conceptual model (Fig. 8.1) should be confirmed (or refuted) and a molecular-biology approach should help to clarify them. Finally, the conceptual model should be transformed into a genuine, mathematized model, in order to be able to make quantitative simulations, to assess orders of magnitude and to confront the model with experimental data.

It remains also to investigate whether it is possible to yield practical applications from the acquired knowledge, especially by manipulating the storage and recall functions. For example, one could try to promote the storage and recall of information leading to economically interesting responses. Conversely, it could be attempted to prevent the storage of undesirable information or, if such information had already been stored, to block its recall and expression.

Epilogue

Arrived at the end of this story, some readers will perhaps feel a little frustrated. From the finding that plants are able to "remember" something, they could dream of esotericism and expect sulphurous revelations, unnatural communication opportunities between plants and us. That is however impossible because animal (including human) logic is quasi opposite to plant logic.

Animals are mobile beings. They can move around for feeding, finding sexual partners and, more generally, searching the best possible environmental conditions. They are afraid of being hurt or killed, which urge them to escape predators. It is thus requisite for them to remember huge information (concerning places, things, time lapses, events, feelings, etc.) in order to make it out in space and time in their trips. To achieve this, they possess neurons and (for the highest species) a central nervous system of tremendous complexity (approximately 100 billions of interconnected neurons in a human brain).

Plants have to stay at the place where they have rooted. They cannot search for optimal environmental conditions. On the contrary they have to adjust their metabolism and growth to the nearby environmental conditions (Fig. 7.1). As far as we can judge, they resent neither pain nor fear. They possess no neurons but have been endowed with a faculty of memory apparently based on a simple interplay between genes. Despite their small capacity, the "learning" and "storage/recall" forms of memory have the potential to decrease/increase the intensity of the response to a stimulus depending on the nature of that stimulus, to produce an integrated response to all the stimuli and stresses to which plants are subjected over time and to synchronize the recall and expression of the memorized information with other internal and external events (including rhythms). In brief, however rudimentary plant memories may be compared to animal memories, they have the potential to play a crucial role in the accommodation of plants to their nearby environment[2].

Is it not there, among many others, an outstanding example of what natural selection can succeed to do!

[2] It would be interesting to investigate if a "storage/recall" form of memory exists or not in fixed animals such as barnacles and corals, and, in this case, if the stored information concerns the stimuli themselves (as do the free animals) or the response to these stimuli (as plants do).

Appendices

Appendix A1 Ciphers and Secret Messages

Life is based on the interactions of coded macromolecules involving either 4-symbol codes (nucleic acids) or 20-symbol codes (proteins). Before explaining these systems (Appendix A.2), we are going to introduce the concepts of "ciphering" code and of correspondence between codes by playing with black or white, geometrically-shaped (circles, squares, diamonds) self-adhesive stickers. The first two games will teach you how to manipulate codes, in the simple case of codes using only four symbols (which will be here the shape and colour of the stickers). The three following games will teach you how to develop and decrypt secret messages, i.e. how to use the stickers for the encoding of messages, so that they are understandable only to who knows the code.

A1.1. The 4-Symbol Codes

First Game Suppose that one has a sheet of four types of different stickers (e.g. black or white squares and circles) and randomly aligns a number of these stickers. An example of the result is what we shall call the "sequence n° 1". The game consists of making another sequence, or "sequence n° 2", which will be deduced logically from the sequence n° 1, using the same types of stickers. This is done by using a "key", i.e. a rule defining the correspondence between the stickers of the two sequences. For example, let the black squares and circles of the first sequence correspond, respectively, to white squares and circles of the second sequence (and vice versa), i.e. (Fig. A1.1).

Starting from any sequence, you can train yourself to build the complementary sequence using the same rule of correspondence, and you will find that this presents no difficulty.

M. Thellier, *Plant Responses to Environmental Stimuli*,
DOI 10.1007/978-94-024-1047-1

Fig. A1.1 Form the complementary sequence of a 4-symbol sequence of stickers, using the same types of stickers in the two sequences

Fig. A1.2 Forming the complementary sequence of a 4-symbol sequence of stickers in the case when one of the types of stickers is not the same in the two sequences

Second Game For this second game, we need to have two sheets of stickers. The first sheet, still containing black or white squares and circles, will serve to build again the sequence n° 1. The second sheet contains white squares and circles as well as black circles, but instead of the black squares, it contains black diamonds. The game consists of building a sequence, or "sequence n° 3", complementary to the sequence n° 1. Again, we need a rule of correspondence. The rule will be, for instance, to continue to associate the white circles of sequence n° 1 with the black circles of sequence n°3 (and vice versa) as well as the black squares of the sequence n° 1 with white squares of sequence n° 3; but we now associate the white squares of sequence n° 1 with black diamonds in sequence n° 3. The result, which is not much more complicated to obtain than that in the first game is given in Fig. A1.2.

Here, again, you may start from any sequence built with the stickers of the first sheet and build its complementary sequence using the stickers of the second sheet.

A1.2. Elaboration and Deciphering of Secret Messages

Third Game We shall use the second sheet of stickers (black and white circles, white squares and black diamonds) to write a secret message, by matching each type of stickers with a letter of the alphabet. *A priori*, we cannot encode more than four letters since we have only four types of stickers. In fact, we cannot even encode more than three letters, since we shall have to keep a type of sticker, termed *stop*, to separate the words in the secret message. For example, adopt the key (i.e. the correspondence between stickers and letters) as indicated below (Fig. A1.3):

There are only two words, C (the Latin numeral for 100) and "cab" that make sense, i.e. (separating them with a *stop*) (Fig. A1.4):

It is frustrating. One cannot build a meaningful sentence with that! To make a good coding with the correspondence one sticker/one letter, we would have had to have as many different stickers as there are letters in the alphabet plus at least one

Fig. A1.3 Assigning a letter or a *stop* instruction to each type of stickers

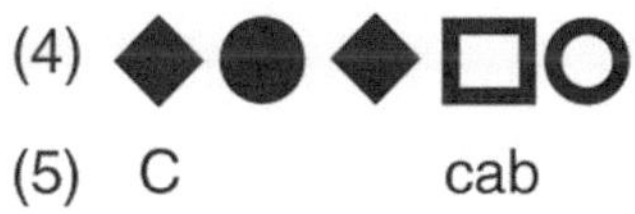

Fig. A1.4 The two words (separated by a *stop*) that can be coded using the one-symbol/one-letter correspondence indicated in Fig. A1.3

Fig. A1.5 Encoding letters and *stop* with pairs of stickers

Fig. A1.6 Decipher a coded message (line 6) into a meaningful sentence (line 7) using the correspondence (2-symbols/1-letter) indicated in Fig. A1.5

for the *stop*, namely, 27 in total. Is it possible to imagine a way to improve the coding efficiency while continuing to use only four types of stickers? If you do not find the solution, the fourth game will put you on the track.

Fourth Game What went wrong in the third game is that we made a one-to-one correspondence (one letter for one sticker type). Now try to match letters to pairs of stickers. For example, adopt the correspondence given in Fig. A1.5.

We see that we can form 16 pairs of stickers, so encode 15 letters (e.g. "a" to "o") and a *stop* instruction. Assume that you receive the secret message that we shall call the "sequence n° 6", and try to find the translation in clear. The correct result is given by the "sequence n° 7" below (Fig. A1.6).

The efficiency of coding by pairs of stickers is already much better than the one-to-one coding used in the previous game. Now, we can encode and decode real messages, such as that from a beggar announcing "I need food". But it is not yet perfect. With this coding system, our beggar, for example, would not have had the opportunity to send a message warning that he needed a drink. He would have missed the letter "r". Well, if pairs of stickers are not enough, let us try to use triplets. This will be the subject of the fifth and last game.

Fig. A1.7 Encoding letters and *stop* with triplets of stickers

Fifth Game Now, let us use triplets of stickers to encode all the letters of the alphabet and establish the list of triplets in a systematic way (Fig. A1.7), as we did for the list of pairs of stickers in the fourth game.

Complete the list of triplets to the end. You will see that there are 64. A mathematician would have been faster by calculating that one can form 4^3 different triplets (which make 64) with the four types of stickers. It is obvious that there are now far too many triplets, since there are only the 26 letters of the alphabet and the *stop* to encode! One can then find various, more or less complicated ways to define the encoding key. For instance, it is possible to:

1. Simply use the first 27 triplets of the list to encode the 26 letters and the *stop*.
2. Use the 64 triplets by matching two triplets with each letter (e.g. by encoding the "a" either by three white squares or by a white circle followed by two black diamonds); 52 triplets would thus be used for the 26 letters of the alphabet, and the *stop* would be coded by either of the 12 remaining triplets.
3. Again, use the 64 triplets by matching a variable number of triplets (for instance, 1–6) with each letter of the alphabet and keeping one or more triplets to encode the *stop*.

We are going to see (Appendix A.2) that life has chosen this last way to proceed.

A1.3. Cryptography

Of course, the manufacture of secret messages (also termed "encrypted messages" or "cryptograms") is not based on the use of codes of stickers! "Cryptography" has become a science, in which teams of mathematicians toil to develop ways to encode messages that are increasingly difficult to decipher. Most often, these sophisticated encodings are intended for commercial, diplomatic or military applications.

To terminate this appendix, let us give rather in the sentimental. A gentleman has an amusing way to use triplets of four types of pearls (e.g. differing in shape, size or colour) to declare his flame to his wife or girlfriend. To do this, he will thread these pearls on a long yarn in order that the sequence of triplets means "*Istoplovestopyoustopichstopliebestopdichstopjestoptstopaimestop*[1] and so on", in as many languages as will be possible. He will tie the two ends of the yarn to make a necklace that he will offer to the woman in his life. Of course, he should not forget to give her the key of encoding. When she will wear this necklace, she will be alone to know what contains this tender cryptogram.

---

[1] I love you, ich liebe dich, je t'aime, etc.

Appendix A2 The Molecular Magic

Entities that we can call "informational macromolecules" or "encoded macromolecules" play an essential role in living beings. A macromolecule is a large molecule, i.e. a molecule formed of a large number of atoms. Encoded macromolecules, including proteins and nucleic acids (DNA and RNA), are long sequences of small molecular units. The order in which these small units are arranged is a code that gives meaning to the macromolecule. It is somewhat similar to the way in which the order of sequence of the 26 letters of the alphabet (plus the spaces and the punctuation) is a representative code of the words in a sentence and gives meaning to the latter. It is also similar to how the stickers have enabled us to develop secret messages (Appendix A.1).

The understanding of the mechanisms involved by encoded macromolecules was the great success of the biologists of the twentieth century, also inspired by famous physicists (Schrödinger 1944; Gamow 1955). We can say that what molecular biology and genetics have allowed for understanding the foundations of life is one of the most beautiful advances that science has ever produced. Detailed explanations can be found in large volumes (Alberts et al. 1983). Here, we shall limit ourselves to a brief introduction, given in the form of answers to 11 questions that the reader may arise.

A2.1. Proteins: What Are They and How Do They Work?

Thousands of substances are present in living matter, and an almost infinite number of chemical reactions can take place *a priori* between them. If all of these reactions had the same probability of occurring, the cell medium would be a completely anarchic reaction system. But the cell possesses catalysts (termed "enzymes") that allow the only reactions that they control to occur at a rate compatible with life. Non-catalysed reactions are so much slower that they can be considered non-existent.

Similarly, the diffusion of solutes between the aqueous extracellular medium and the cytosol (which is also aqueous) is strongly limited by the lipid nature of the membrane (the so-called plasma membrane) that separates them. Again, specific catalysts, which are implanted in the membrane, allow only the exchanges of substances that they control to occur at a rate compatible with life. These catalysts of transmembrane exchanges are called "carriers" (or "ion channels" in the particular case of the transmembrane transport of some mineral ions[2]).

---

[2] Ions are molecules bearing one or several elementary electrical charges (the "elementary charge" is equal in absolute value to the charge of the electron). Ions exist that bear a negative charge (the so-called anions) and others bearing a positive charge (the so-called cations). There are mineral ions (such as Cl^-, Na^+, Ca^{2+}, etc.) and organic ions (such as CH_3COO^-, $CH_3NH_3^+$, etc.). The "monovalent" ions are those bearing a single elementary charge (e.g. CH_3COO^- or Na^+), the "divalent" ions are those bearing a double charge (e.g. Ca^{2+}), etc. When they are dissolved in water, salts,

Enzymes, carriers and ion channels are all proteinaceous, among a variety of other proteins involved in all aspects of the functioning (and even of the structure) of cells.

Therefore, in living cells, proteins play a role somewhat equivalent to the role played by tools in a workshop. They are often very specific with regard to both the processes that they control and the substances (or "substrates") involved in these processes. They are made of sequences of a few dozen to a few hundred "amino acids" (which are small molecules bearing at least one acid and one amine function). Chemists are aware of the existence of a large number of amino acids, but only 20 (listed in Sect. A.2.6) participate in the constitution of the proteins found in living systems. Proteins fold in space in a complex manner, thus forming "active sites". For example, in the case of enzymes, the "catalytic site" is an active site to which the substrate binds and undergoes its transformation. However, it is the order of sequence of the amino acids (the so-called primary structure of the protein) that imposes the spatial structure of the protein and governs its function and its specificity.

A2.2. What Is a Gene?

The concept of "gene" existed long before molecular genetics. At the beginning, a gene was the abstract entity that encoded a hereditary characteristic (most often the primary structure of one or several proteins) in the cell nucleus. At that time, no one knew what this entity was exactly (i.e. what was its molecular nature). It was just known that it had a real physical existence, especially because geneticists had developed an ingenious method to identify the relative positions of the different genes with one another. For the modern interpretation of the gene, see Sect. A.2.3.

A2.3. What Is DNA, and What Is its Structure?

DNA (Fig. A2.1a) is a macromolecule made of a double strand of four types of "deoxyribonucleotides" (a particular case of the chemical species termed "nucleotide"). Here, we shall not enter in the detail of the chemical formulae of the nucleotides. We shall simply characterize these nucleotides using their classical symbols (A, C, G and T). In each of the DNA strands, the order of the nucleotide sequence is a 4-letter code (resembling the 4-sticker code in the first game of Sect. A.1.1). The name of DNA (for "*d*eoxyribo*n*ucleic *a*cid") comes from the fact that its constitutive nucleotides contain "deoxyribose" (a sugar that is not very well known by non-specialists).

---

acids and bases split (in total or in part) into anions and cations. For instance, in water, the sodium chloride (NaCl) dissociates into Na^+ and Cl^-, the hydrochloric acid (HCl) into H^+ and Cl^- and the caustic soda (NaOH) into Na^+ and OH^-. There are macromolecules or multimolecular systems that bear a large number of elementary charges; they can be characterized by their charge density (e.g. the mean charge per length unit for unidimensional systems).

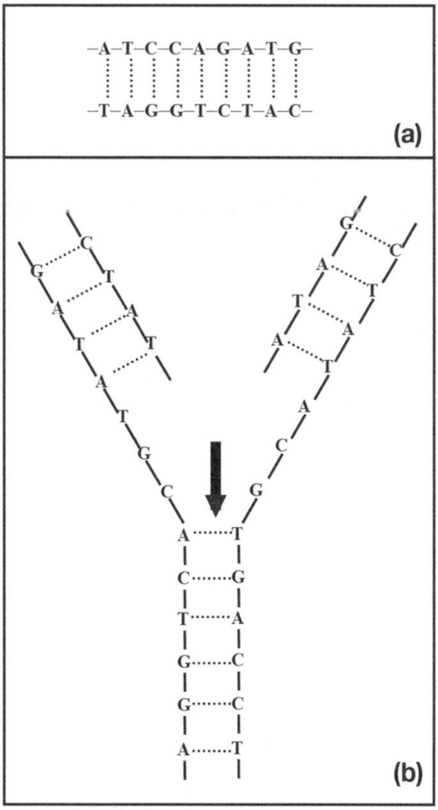

Fig. A2.1 *Schematic representation of DNA and of its replication.* (**a**) Diagram of the double strand of DNA with its 4-letter code (A, C, G, T), the letters being arranged so that an A in one strand is always facing a T in the other strand, and a C in one strand is always facing a G in the other strand. In reality, the double strand is not linear but organized into a double helix (not shown in the figure). (**b**) To duplicate DNA, a first enzyme system (*thick arrow*) separates the two strands from each other, and then a second enzyme system reconstructs a new strand in front of each of the initial, separated strands according to the complementarity A/T and C/G. This results in getting two double strands that are identical with each other and identical with the initial double strand of DNA. Symbols: *full lines* = covalent bonds, *dotted lines* = hydrogen bonds

The two strands of DNA are arranged in a way such that there is always an A in front of a T and a C in front of a G, and vice versa (Fig. A2.1a). In each strand, the nucleotides are connected, each to the next, by strong "covalent" bonds. By contrast, the bonds between T and A or between G and C (termed "hydrogen bonds"), which tie the two strands with each other, are much weaker than covalent bonds. The two strands thus are complementary as were the sticker sequences n° 1 and 2 in the first game of Sect. A.1.1. DNA molecules are up to several metres long! They are arranged in space in a double helix. Most of the DNA of a cell is located in the cell nucleus in a folded and compacted form. In what follows, we shall mainly take interest in this nuclear DNA. However, there are also small amounts of DNA in mitochondria and chloroplasts (Sect. A.2.10).

Nowadays, it is known that the hardware support of genes is the macromolecule of DNA or, more precisely, a small fraction of the DNA molecule termed "coding DNA". More specifically, a gene consists of a sequence of nucleotides in a coding fraction of one of the DNA strands. This sequence is a pattern coding for the primary structure of a protein. DNA thus is the genetic memory of the cell. Except when a mutation occurs (Sect. A.2.8), nuclear DNA is transmitted without any modification in the process of cell division, and it is the same in all the cells of a given organism. The exact role of the non-coding DNA is still not clearly understood. It may be that it is mainly involved in regulatory processes. Sometimes, it occurs that a given gene codes for several different proteins or that the participation of several genes is requisite for the coding of a single protein.

A2.4. How Does DNA Keep Identical to Itself in Cell Division?

It is the fact that the two DNA strands are complementary with each other that allows DNA to be kept identical to itself in the process of cell division. Prior to entering into mitosis,[3] a cell "replicates" (or "duplicates") its nuclear DNA (Fig. A2.1b). To do this, a first enzyme system breaks the hydrogen bonds, thus separating the two DNA strands from each other somewhat in the way you open a zipper. Then another enzyme system restores a second strand in front of each of the two separated strands with matching A to T, C to G, G to C and T to A. The Fig. A2.1b shows that (again except accident) this process yields two double-strands identical with each other and identical with the initial double strand. Each of the two "daughter cells"[4] obtained by the division of the initial cell (also termed the "mother" cell) carries one of the double strands. The daughter cells thus contain exactly the same nuclear DNA, and this nuclear DNA is identical with that of the mother cell. Since the cells of any organism (e.g. a plant) come from a single initial cell by successive divisions, they all carry (again except accident) exactly the same nuclear DNA.

A2.5. What Is RNA, and What Is its Structure?

RNA (for *r*ibo*n*ucleic *a*cid) molecules are made of a single strand constituted of four types of nucleotides (termed "ribonucleotides" because the sugar ribose replaces the DNA deoxyribose). Three of these nucleotides resemble those in DNA (except for the sugar) and are represented by the same symbols (A, C and G), while

[3] Mitosis is a "normal" cell division, i.e. other than those leading to the formation of reproductive cells.

[4] Note that the words "mother" and "daughter", as we are using them here, do not have their usual meaning. In everyday life, a mother that has produced daughters remains as she was before. Here, the mother cell splits into two daughter cells and thus does not exist any longer after the division.

the fourth one, quite different from the deoxyribonucleotide T, is represented by the symbol U. In RNA molecules, the order of the nucleotide sequence is again a 4-letter code. RNA strands are considerably shorter than DNA molecules. Various types of RNA molecules exist, for example, soluble RNA (or sRNA), ribosomal RNA (or rRNA) entering in the composition of small cytoplasmic organelles involved in the biosynthesis of proteins and messenger RNA (or mRNA) the role of which will be discussed in Sect. A.2.6.

A2.6. How Are Proteins Biosynthesized on the Pattern of DNA?

Consider a nuclear gene represented by its nucleotide sequence on one of the DNA strands. This sequence is within the nucleus, while the metabolic machinery requisite for protein synthesis is within the cytoplasm. Hence, the synthesis of the protein corresponding to that gene has to arise in two steps. The first step (termed "transcription") consists in building, on the pattern of the gene DNA an mRNA that will migrate to the cytoplasm. Transcription therefore serves to transfer the information brought by the gene from the nucleus to the cytoplasm. The second step (termed "translation") concerns the actual production of a protein on the pattern of the mRNA.

To transcribe the DNA pattern (Fig. A2.2a) into a pattern of mRNA, enzyme mechanisms (1) separate the two strands of DNA at the level of the site coding for the protein under consideration (Fig. A2.2b); (2) position the free ribonucleotides A, C, G and U so that A is in front of T, C in front of G, G in front of C and U in front of A (Fig. A2.2c); (3) bind these nucleotides from one to the next by covalent bonds, thus producing an mRNA molecule complementary of the coding part of the DNA; and (4) release this mRNA (Fig. A2.2d) that can then migrate from the cell nucleus to the cytoplasm. Transcription does not present any particular difficulty since it comes to establish the correspondence between two 4-letter codes that (A, C, G and T) of DNA and that (U, G, C and A) of the mRNA. This is the same problem as that which has been solved in the second game of Sect. A.1.1.

Translation consists in producing a protein on the pattern of mRNA. This is equivalent to matching a 20-letter code (the 20 amino acids entering in the constitution of proteins) with the 4-letter codes of DNA and mRNA. It is obvious that a letter-to-letter correspondence is not suitable. Let us reason as was done in the third, fourth and fifth games of Sect. A.1.2. Considering all the pairs of DNA letters (TT, TG, TC, TA, GT, GG, etc.), or of mRNA letters (AA, AC, AG, AU, CA, CC, etc.), that it is possible to write down with four letters, it is easily calculated that there are only 16 of them, which is insufficient to encode the 20 amino acids of proteins. The most economical solution is therefore to use triplets of letters of nucleic acids for coding the amino acids of proteins. Life has adopted this solution; however, there are now many more triplets (64) than the 20 amino acids of proteins! To solve the difficulty, natural selection has retained the most complicated possibility (solution (3) of the fifth game in Sect. A.1.2), which consists of encoding the *stop* and some amino acids with several different triplets. Table A2.1 gives the correspondence

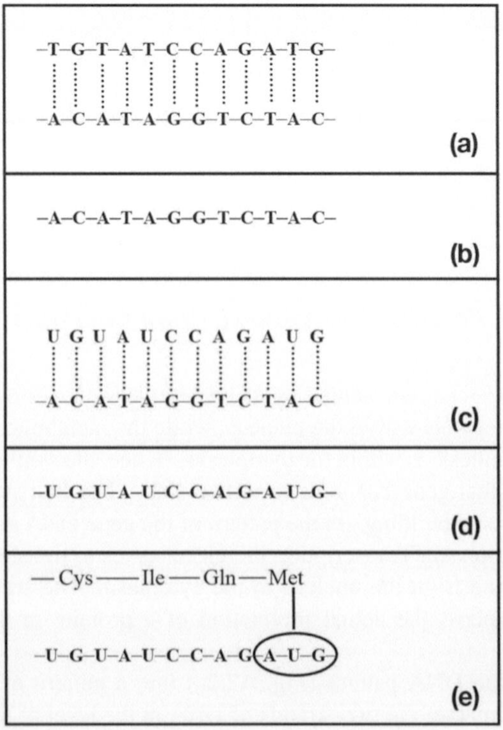

Fig. A2.2 Protein biosynthesis: DNA transcription into mRNA and mRNA translation into proteins. (**a**) The double strand of DNA. (**b**) A single strand of DNA obtained by the local breaking of the hydrogen bonds which were responsible for the stability of the double strand. This single strand of DNA serves as a template for its transcription into the corresponding mRNA molecule. (**c**) Positioning nucleotides in the appropriate order to form the mRNA to come. (**d**) Release of the mRNA after the establishment of covalent bonds between its constituent nucleotides and the breaking of the hydrogen bonds between DNA and mRNA. (**e**) Translation of the mRNA into a protein by a ribosome (here symbolized by an *ellipse*) according to the genetic code (Table A2.1)

between the DNA or mRNA triplets (often called "codons") and the protein amino acids. This table is termed the "genetic code". There may be up to six different codons for one amino acid (e.g. it is the case with serine), while there is only a single codon for methionine. As for the *stop* codons, they are at the end of all genes. In cases in which a single mRNA codes for several different proteins, they are used to indicate that the code of the first protein is ended and that starts the second code, etc. To reflect the fact that several different codons can code the same amino acid, the genetic code is said to be "degenerate".

Translation is carried out through the intervention of ribosomes (Sect. A.2.5). They are small organelles formed by the association of some rRNA molecules with several proteins. Ribosomes slip along the mRNA thread where they play a role somewhat equivalent to that of a sound head reading a magnetic tape. They progress from one codon to the next. At the level of each codon, they position the amino acid

Table A2.1 The genetic code: correspondence of the DNA and mRNA codons with the constituent amino acids of proteins

Codon	Codon	Amino acid		Codon	Codon	Amino acid	
(DNA)	(mRNA)	Name	Symbol	(DNA)	(mRNA)	Name	Symbol
TTT	AAA	Lysine	Lys	CTT	GAA	Glutamic acid	Glu
TTG	AAC	Asparagine	Asn	CTG	GAC	Aspartic acid	Asn
TTC	AAG	Lysine	Lys	CTC	GAG	Glutamic acid	Glu
TTA	AAU	Asparagine	Asn	CTA	GAU	Aspartic acid	Asn
TGT	ACA	Threonine	Thr	CGT	GCA	Alanine	Ala
TGG	ACC	Threonine	Thr	CGG	GCC	Alanine	Ala
TGC	ACG	Threonine	Thr	CGC	GCG	Alanine	Ala
TGA	ACU	Threonine	Thr	CGA	GCU	Alanine	Ala
TCT	AGA	Arginine	Arg	CCT	GGA	Glycine	Gly
TCG	AGC	Serine	Ser	CCG	GGC	Glycine	Gly
TCC	AGG	Arginine	Arg	CCC	GGG	Glycine	Gly
TCA	AGU	Serine	Ser	CCA	GGU	Glycine	Gly
TAT	AUA	Isoleucine	Ile	CAT	GUA	Valine	Val
TAG	AUC	Isoleucine	Ile	CAG	GUC	Valine	Val
TAC	AUG	Methionine	Met	CAC	GUG	Valine	Val
TAA	AUU	Isoleucine	Ile	CAA	GUU	Valine	Val
GTT	CAA	Glutamine	Gln	ATT	UAA	*stop*	–
GTG	CAC	Histidine	His	ATG	UAC	Tyrosine	Tyr
GTC	CAG	Glutamine	Gln	ATC	UAG	*stop*	–
GTA	CAU	Histidine	His	ATA	UAU	Tyrosine	Tyr
GGT	CCA	Proline	Pro	AGT	UCA	Serine	Ser
GGG	CCC	Proline	Pro	AGG	UCC	Serine	Ser
GGC	CCG	Proline	Pro	AGC	UCG	Serine	Ser
GGA	CCU	Proline	Pro	AGA	UCU	Serine	Ser
GCT	CGA	Arginine	Arg	ACT	UGA	*stop*	–
GCG	CGC	Arginine	Arg	ACG	UGC	Cysteine	Cys
GCC	CGG	Arginine	Arg	ACC	UGG	Tryptophan	Try
GCA	CGU	Arginine	Arg	ACA	UGU	Cysteine	Cys
GAT	CUA	Leucine	Leu	AAT	UUA	Leucine	Leu
GAG	CUC	Leucine	Leu	AAG	UUC	Phenylalanine	Phe
GAC	CUG	Leucine	Leu	AAC	UUG	Leucine	Leu
GAA	CUU	Leucine	Leu	AAA	UUU	Phenylalanine	Phe

corresponding to this codon (Fig. A2.2e) and link the amino acids so arranged in the order imposed by the mRNA pattern. When a ribosome is sufficiently advanced on an mRNA, a second ribosome comes to position itself on the same mRNA, then a third when the second is sufficiently advanced and so forth. An mRNA thus carrying a series of ribosomes reading the mRNA and synthesizing the corresponding protein is called a “polyribosome” or, more simply, a polysome.

A2.7. How Does Cell Differentiation Occur?

All the cells of an organism have the same sequences of nucleotides in their DNA, and yet they have neither all the same shape nor the same metabolic behaviour. They are said to have differentiated into various sorts of cells (e.g. epidermal, parenchymal or phloem cells, etc.).

During cell differentiation, a certain number of genes lose their ability to be transcribed: we say that they get "locked". The processes (termed "epigenetic" processes) which are involved in gene locking/unlocking are complicated. Some sorts of proteins (termed "histones") and sRNA are implicated, and reactions (such as methylation/demethylation or acetylation/deacetylation) also take place. Experimentally, in *Arabidopsis* plants, it was found that notable changes in the methylation status of genes (and therefore changes in the relative numbers of locked/unlocked genes) accompanied the development of the plants (Gehring et al. 2009). Demethylation reactions can also occur during the formation of tissues containing finally poorly methylated DNA (see, e.g. Thellier and Lüttge 2013 for references). Eventually, in the cells of each sort of tissue, only the genes coding for the proteins useful in this tissue remain unlocked.

It may happen that locked genes get unlocked under the effect of diverse internal or external events and resume their activity. Differentiated cells then tend to dedifferentiate. Inversely, the epigenetic state of a gene or of a group of genes may be so firmly established that it is heritable through successive cell divisions.

An unlocked gene is not necessarily an active gene (i.e. a gene that is transcribed). In some cases, an unlocked gene will be effectively active only in the presence of specific substances (termed "activators") and in the absence of other substances (termed "repressors").

The locking and unlocking of genes are usually robust operations, not easily reversible. By contrast, the activation of an unlocked gene requires only the presence of a suitable activator and the absence of any repressor (and inversely the repression of an unlocked gene requires the presence of a repressor and/or the absence of activators).

A2.8. How Can the Evolution of Living Beings Be Explained?

We have seen above that, in principle, DNA keeps identical to itself during replication (Sect. A.2.4), and yet living beings evolve over the course of millennia. The explanation is that DNA replication is not always perfect. It happens that errors (termed "mutations") occur, e.g. that a G instead of a T gets positioned in front of an A or that part of the replication fails, etc. Given how replication occurs (Fig. A2.1b), the error is perpetuated in one of the two daughter cells after the first cell division and then in all the cells which derive by division of this daughter cell. These errors are rare; moreover, there are DNA repair mechanisms that still limit their occurrence. However, the frequency of such errors, while being very low, is not strictly zero.

Fig. A2.3 Example of a somatic chimera on a spindle tree (Euonymus japonicus) (Photo: Marie-Claire Verdus). The plant bears usually *yellow-edged* leaves. The leaves carrying the mutation appear *white edged*

When a mutation affects a somatic cell (i.e. other than a reproductive cell), it has effect only in that cell and in those deriving from it by division. This may be the origin of "chimeras" (Fig. A2.3). When a mutation touches a germ cell, this can affect the descendants of the organism. There are mutations that entail no consequences. They are said to be "silent". Most often, silent mutations occur in non-coding regions of DNA. Silent mutations may also happen in a coding region, when one triplet is replaced by another one coding for the same amino acid; for example, it is visible in Table A2.1 that, if a mutation changes a triplet TGT of DNA into TGC, this will continue to code for threonine. Most of the non-silent mutations are unfavourable or even lethal; but it also happens that a mutation is beneficial, and that it is retained by natural selection, thus contributing to the evolution.

In short, the replication of DNA is biologically satisfactory (i.e. allowing both the permanence of species on a sufficiently short period and their evolution over much longer periods) only because it makes mistakes but infrequent mistakes.

A2.9. Is the Genetic Code Dependent on the Species?

Apart for minor variations observed with specific organisms (Baroin Tourancheau et al. 1995), the genetic code is universal, and the properties and functioning of DNA and mRNA, such as described above in plants, are the same throughout the living world. Under its apparent diversity, life thus presents an astounding unity of conception. Does this mean that life arose only once, and that the organization of the first cell was kept, almost without changes, in all beings that have derived from this initial cell? Or, more simply, does this mean that the organization of life, as we know it, was so efficient that it has outperformed and eliminated all other attempts to organization that nature could do? We shall not try to answer that. In future, if space research allows finding other forms of life than ours, it will be time to come back to these questions.

A2.10. What About the Mitochondrial and Chloroplast DNA?

It has been reported above (Sect. A.2.3) that mitochondria and chloroplasts contain a small amount of DNA. This mitochondrial and chloroplast DNA looks more like bacterial DNA than like the nuclear DNA of plants. It is believed that mitochondria and plastids are derived from bacterial symbionts that eventually integrated into the cells. Specific studies of sequence comparison have demonstrated that these organelles actually originate from bacteria quasi similar to still existing bacteria. The bacterial families to which they belong have even been elucidated. Whatever it is, mitochondrial and plastid DNA code for a few proteins or fragments of proteins. There are indeed proteins that are encoded by mitochondrial or chloroplast DNA only for a part, the rest of the protein being encoded by nuclear DNA.

In the fertilized egg and the organism deriving from it, the nuclear DNA originates in equal parts from the male and female gametes. However, in the fertilized egg, the mitochondria and plastids originate only from the female gamete. As a consequence, the mitochondrial and chloroplast DNA comes only from the ovum.

A2.11. What Is a GMO?

The acronym GMO stands for "genetically modified organism". In the usual acceptation, a GMO is an organism in the DNA of which one or several exogenous genes (more precisely exogenous DNA fragments) have been inserted, under such conditions that this/these exogenous gene(s) is/are transcribed and translated (Sect. A.2.6). The process through which exogenous genes are inserted in the genome of an organism is termed "transgenesis", and the inserted genes are termed transgenes.

Rigorously speaking, almost all of our food comes from genetically modified organisms, in the sense that they are organisms the genome of which has been modified from that of natural species by long-term agriculture and farming practices. This means that most of the organisms that we eat have been obtained by a slow process of selection over many generations. These organisms are often termed GIO (for "genetically improved organisms") rather than GMO. Transgenesis is considerably faster and more directive. In the current state of the art, it is still not possible to incorporate transgenes at will into a specific site in the DNA of the recipient organism. Transgenes are therefore incorporated haphazardly into the receiving DNA. It may occur that a transgene changes its location in the course of the successive generations or after a transfer (intended or due to contamination), which can alter its function. Thus, transgenesis presents interesting possibilities, which are already used, for example for basic research (Box 2.4) and for pharmacology, but it can also present risks that have to be studied.

Appendix A3 Calcium Condensation/Decondensation

We are going to briefly describe the condensation/decondensation of ions (Manning 1969), especially of the Ca^{2+} ion. Often ignored in manuals, it may yet play a switch role in biological processes as seen in our attempt to model plant memories (Sect. 8.2.2).

A3.1. Water Condensation

A sufficiently cold object that is introduced in a gaseous atmosphere containing water vapour gets immediately covered with tiny drops of water. It is said that the water vapour has "condensed" on the object. This is what happens when your glasses or the windscreen of your car are covered with moisture. Water molecules, which were diffusing freely in the atmosphere, are literally collapsing on the cold object. This condensation of water presents two remarkable properties: (1) it is an all or nothing phenomenon and (2) if the water molecules can continue to diffuse freely in the droplets of moisture, they can no longer get away from the object on which they are condensed. Furthermore, water condensation is reversible. If you heat up the object, the moisture that was covering it reverts to the vapour state, and the water molecules become free to again disseminate in the atmosphere. This is what permits the defrosting of the windows of your car.

A3.2. Ionic Condensation

Now consider (Fig. A3.1) fixed (i.e. non-diffusible) one-dimensional structures (termed "1-D structures"),[5] loaded with negative electrical charges and dipped in a solution (or a gel) containing free negative and positive ions (including the divalent

[5]A "fixed structure" is a molecule or a multimolecular structure large enough to remain virtually motionless in the cell medium. By contrast, small ions and small solutes that can diffuse within the

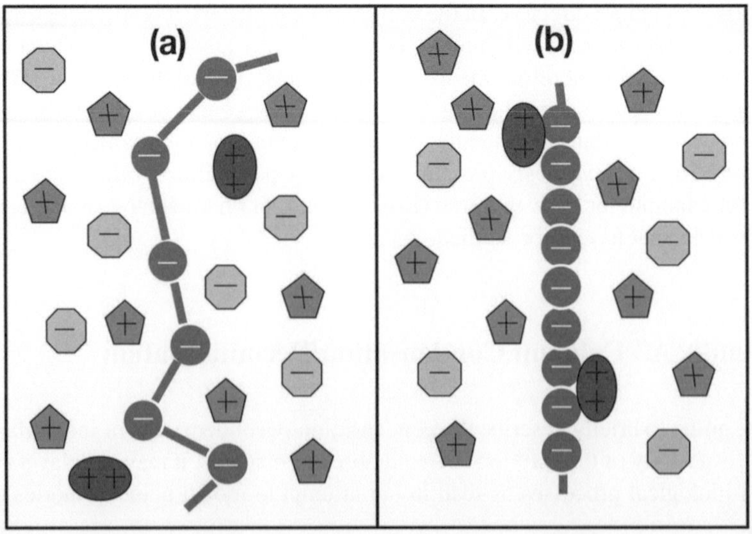

Fig. A3.1 Calcium condensation on a fixed 1-D structure loaded with negative electric charges (From Thellier and Lüttge 2013). (**a**) When the density of negative charges of the fixed 1-D structure is low (less than a critical value, ξ_c), small mono- and divalent ions diffuse freely in the bathing solution. (**b**) When the density of negative charges of the fixed 1-D structure becomes higher than ξ_c, small divalent ions (here Ca^{2+}) "condense" on the fixed structure. The local electrical neutrality is statistically respected, i.e. at any location within the system, the number of negative charges is always very close to that of the positive charges. The fixed 1-D structure may have a shape more or less tortuous when its charge density is low [case (**a**)]; but, under the effect of the electrical repulsion, it tends to straighten when its charge density increases [case (**b**)]

calcium ion, Ca^{2+}). Such structures are frequently found in cellular media (e.g. nucleic acids, linear associations of proteins, constituents of the cytoskeleton, etc.).

The density of negative charges on a fixed 1-D structure is conventionally represented by the Greek letter ξ (pronounced "ksi"). When the ξ value is sufficiently low, the mobile ions can diffuse freely in the bathing solution (or gel). However, if some event causes ξ to increase to a value greater than a critical value (conventionally written ξ_c), the free Ca^{2+} ions will collapse on the fixed 1-D structure so that ξ tends to decrease towards the critical value, ξ_c. It is said that they "condense" on the fixed 1-D structure. Ca^{2+} condensation presents two remarkable properties: (1) it is an all or nothing process (i.e. a process which, unlike the usual chemical processes, does not obey the law of mass action) and (2) when Ca^{2+} ions are condensed on a fixed 1-D structure, they continue to move along this structure, but they can no longer go away from it. The ionic condensation is reversible. If some event causes ξ to become less than ξ_c, some calcium will "decondense" (i.e. will come back to the form of free Ca^{2+} ions) so that the ξ value tends to increase towards the critical value, ξ_c (Manning 1969).

cell medium are said to be "mobile" (or "free"). A 1-D structure is a structure with a single dimension or, at least, with a main dimension. For example, at our scale, a dressmaker thread (whether stretched or more or less twisted) is a 1-D structure; at the molecular scale, DNA or RNA molecules, filaments and tubules of the cytoskeleton, etc. are 1-D structures.

There is an obvious similarity in the behaviour of a water vapour tending to condense on a cold object and that of free calcium ions tending to collapse on a 1-D structure with a sufficiently high density of negative charges. This is the reason why the same word, "condensation", is used for these two processes. That said, it is clear that the condensation of water vapour and ionic condensation are not one and the same thing. The first can occur in a gaseous atmosphere on surfaces (which are two-dimensional structures) with temperature as the main variable, while the second occurs only on one-dimensional structures in an aqueous solution or gel with the charge density being the main variable. In fact, the familiar phenomenon of water condensation is a convenient image, helping to intuitively represent the more mysterious phenomenon that is the condensation of Ca^{2+} ions.

A3.3. Biological Implications

There are many cellular systems the activity of which depends on the presence of Ca^{2+} ions. For example, this is the case with a certain number of enzymes (termed "calcium-dependent enzymes") that are practically inactive in the absence of calcium but become active when the Ca^{2+} concentration is increased. Assume that some of these Ca^{2+}-dependent enzymes are attached to negatively charged, fixed 1-D structures, while others are scattered in the surrounding medium. When calcium is not condensed (low ξ value), its concentration close the fixed structure is approximately the same as it is in the surrounding medium, and the activity of both types of enzymes depends only on the mean concentration of calcium. When calcium condenses on the fixed structure (ξ getting larger than ξ_c), the concentration of calcium at the level of the fixed structure is very much increased, while that in the surrounding medium is very much decreased. As a consequence, the activity of the enzymes attached to the fixed structure is very much increased, while the enzymes scattered in the surrounding medium are inactivated.

When some internal or external event causes the ξ value to become higher (if it was low) or lower (if it was high) than ξ_c, the condensation/decondensation process will act as a switch abruptly changing the relative activity of the enzymes bound to the fixed charges compared to that of the enzymes scattered in the surrounding medium. This may eventually reorient the cell metabolism.

Glossary[6]

Abiotic By contrast with stimuli such as an attack by fungi, bacteria and viruses or wounding by herbivores, an abiotic stimulus is a stimulus that has not been caused by a living being (e.g. wind, rain, thermic shock, etc.).

Aequorin Protein found in the bioluminescent* jellyfish *Aequorea victoria*. In the presence of calcium, this protein allows transforming into light a part of the metabolic energy of the jellyfish.

Alternate See leaves*.

Amino acid Small molecule carrying an acid function (–COOH) and an amine function ($-NH_2$). The general form of an amino acid is $HOOC–R–NH_2$ where R has a more or less complicated form. For example, R takes the form $H–C–CH_3$ or $H–C–CH_2–CH_2–S–CH_3$ for alanine and methionine. Chemistry teaches us that an acid can react with an amine to form an amide. The existence of both an acid and an amine function thus enables two amino acids to bind to each other (with loss of water) to give a "dipeptide" $HOOC–R1–NH–CO–R2–NH_2$ (where R1 and R2 are characteristic of the two amino acids involved). The dipeptide has in turn an acid and an amine function. Therefore, it can bind to a third amino acid (with loss of water) and so on until forming longer or shorter sequences. Proteins are sequences of a few dozens to a few hundred amino acids.

Anion Negative ion*

Apex. A synonym for "terminal bud".

Apical dominance Dominance of the terminal bud (or apex*) on the underlying buds, thus preventing them from growing.

Apoptosis Programmed cell death. By contrast with necrosis, which is a passive consequence of some harmful event, apoptosis is the controlled "suicide" of some cells involved for instance in the normal development of multicellular organisms or in defence reactions against aggressions.

---

[6] Note: the words labelled with an asterisk can be found in this glossary.

M. Thellier, *Plant Responses to Environmental Stimuli*,
DOI 10.1007/978-94-024-1047-1

Atom Particle constituting the smallest conceivable amount of a chemical element. An atom consists of a nucleus, itself made up of two kinds of elementary particles, protons* and neutrons*. Electrons* orbit the nucleus. In the non-ionized atom (see ions*), the number of electrons orbiting the nucleus is equal to the number of protons in the nucleus. This number (also termed the "charge number") characterizes the element under consideration, this element being conventionally represented by its chemical symbol. For instance, hydrogen (chemical symbol H) contains 1 proton in its nucleus and 1 electron orbiting the nucleus. Helium (chemical symbol He) contains 2 protons in its nucleus and 2 electrons. Lithium (chemical symbol Li) contains 3 protons in its nucleus and 3 electrons and so on. All the existing chemical elements are listed in Mendeleyev's table.

Auxin Also termed "indole-acetic acid", this plant hormone is involved in a variety of processes including growth, morphogenesis (formation of new organs), reproduction, response to various stimuli, etc.

Bioluminescence Property shared by various living species that divert a portion of their metabolic energy for the emission of light.

Bivalent A bivalent ion* is an ion carrying a double elementary electrical charge*.

Cation Positive ion*.

Cell The elementary unit of living beings. For detail concerning plant cells, see Sect. 1.3.

Cell dedifferentiation. See cell differentiation*.

Cell differentiation. In an organism, embryonic cells are all similar with one another. Then they progressively get specialized in the different functions characteristic of that organism. For instance, the initially undifferentiated cells of a plant embryo evolve and become parenchymal, epidermal, stomatal, etc. cells*. This is due to the locking of certain genes (for detail, see Sect. A.2.7). In plants it is relatively common that differentiated cells dedifferentiate and return to a state close to the embryonic state.

Cell wall See cell*.

Chlorophyll A green pigment whereby green plants perform photosynthesis*.

Chloroplast A cell organelle containing chlorophyll.

Chromoprotein A molecule made of a protein associated with a pigment (i.e. a coloured substance).

Chromosome During cell division, DNA associated with proteins gets arranged into components, visible by optical microscopy, that are called "chromosomes".

Circadian A rhythm is said to be circadian when its period is close to 24 h.

CNRS Centre National de la Recherche Scientifique [National Centre for Scientific Research].

Codons In the DNA or mRNA strands, codons are triplets of nucleotides coding for an amino acid. For example, the DNA codon TCT, as well as the AGA codon of the corresponding mRNA, encode the amino acid "arginine" during protein biosynthesis (for detail, see Sect. A.2.6).

Cotyledons They are the first leaves of plants. They pre-exist in the seed and widen and turn green after germination. There are monocotyledonous and dicotyledonous plants that have, respectively, a single or two cotyledons.

Cotyledonary bud Bud located at the axil of a cotyledon. In a case of apical dominance*, the cotyledonary buds remain quiescent, and they are so small that they are hardly visible with the naked eye.

Cyanobacteria Photosynthetic bacteria formerly called "blue algae".

Cytoplasm In typical plant cells, the cytoplasm is one of the main cell compartments, located between the wall* and the vacuole(s)*.

Cytosol In the cytoplasm, the cytosol is a complicated aqueous medium within which bathe the nucleus and organelles such as mitochondria, plastids, etc.

Cytoskeleton Network of microfilaments and microtubules that structures the cytoplasm of cells.

Decapitation The decapitation of a young plant is the removal of its terminal bud. It is said that a young plant has been decapitated when its terminal bud has been removed.

Dicotyledonous Higher plants the seeds of which contain two cotyledons. Among the plants mentioned in this book, *Arabidopsis, Bidens, Xanthium*, flax, radish, tobacco, tomato are dicotyledonous plants.

Dioecious A dioecious plant has distinct male and female individual organisms.

Dioecy The characteristic of a plant to be dioecious*.

Dionaea See Venus flytrap*.

Divalent ion Ion* carrying two elementary electrical charges*. For instance, Ca^{2+} is a divalent cation*.

DNA This acronym means deoxyribonucleic acid. It is a long molecule carrying the genetic information of the cells.

Duplication, to duplicate See replication, to replicate*.

Electron Light elementary particle, carrying a negative electrical charge, which orbits the nucleus of atoms.

Elementary electrical charge The negative elementary electrical charge is equal to that carried by an electron*. The absolute value of the positive elementary electrical charge is equal to that of the electron*, but it is a positive instead of a negative charge.

Epigenetic, epigenetics The sequence of nucleotides in the DNA* strands constitutes the genetic memory of a cell. In differentiated cells some of the genes are locked (they are not transcribed*), and only the genes coding for the proteins indispensable to the functioning of the differentiated cell are unlocked (they can be transcribed and translated*). The processes permitting gene locking/unlocking (DNA methylation/demethylation, involvement of histone proteins or of various sRNA) are said to be "epigenetic". Epigenetics is the science studying the processes involved in cell differentiation/dedifferentiation. Note that some epigenetic features can be maintained through cell division. Hereditary characters thus may depend both on genetics and epigenetics. For detail, see Sect. A.2.7.

Familiarization Process by which the repetition of a given, non-dangerous stimulus causes the intensity of the response to another occurrence of the same stimulus to be decreased or annulled.

Fixed 1-D structure A fixed structure is a molecule or a molecular assembly bulky enough so that it does not move within the cell medium. A 1-D structure is a one-dimensional structure, or, at least, a structure that has one main dimension, the other dimensions being considered quasi negligible. At our scale a dressmaker thread is a 1-D structure. At the molecular scale, DNA* and RNA* molecules, the constituents of the cytoskeleton*, etc. are 1-D structures. A fixed 1-D structure is both a fixed and a 1-D structure.

Gamete Reproductive cell, called pollen grain (male cell) and ovum* (female cell) in plants. Fertilization is the union of a male with a female gamete thus forming a zygote*.

Gene Basic physical and functional unit of heredity. Genes are made up of DNA*. For more detail, see Sects. A.2.2 and A.2.3.

Gene locking An active gene* is said to become locked when it loses its ability to be transcribed under the effect of chemical reactions (e.g. methylation) or the presence of certain proteins (e.g. histones) or sRNA.

Gene unlocking A locked gene* is said to unlock if it becomes able to be transcribed again as a result of chemical reactions or of the modification or departure of certain proteins or sRNA.

Genetic code Table of correspondence between amino acids* and DNA or mRNA codons* in protein biosynthesis (see Table A2.1 in Sect. A.2.6).

GMO Genetically modified organism. See Sect. A.2.11.

Gravitropism Growth response of plants to gravity, causing the aerial part to grow upwards (negative gravitropism) and the roots to grow downwards (positive gravitropism).

Hyper-osmotic A plant is said to undergo a hyper-osmotic stress when it is transported from a usual nutrient solution to a nutrient solution added with a high concentration of a substance that the plant cannot absorb. A plant subjected to hyper-osmotic stress tends to lose water in favour of the concentrated solution. A hyper-osmotic stress thus mimics the effect of drought.

Hypocotyl Part of the stem between the root and the cotyledon(s)*.

Ion An ion is an atom or a molecule electrically charged following a loss (positive ion) or a gain (negative ion) of electron(s)*. Positive or negative ions are called cations* or anions*, respectively. Monovalent, bivalent, trivalent, etc. ions are atoms or molecules that have lost or gained one, two, three, etc. electrons. For instance, the bivalent cation Ca^{2+} is an atom of calcium that has lost two electrons.

Isomer A molecule is an isomer of another molecule if the two molecules contain the same numbers of their constitutive atoms but differ in the spatial arrangement of these atoms. For instance, the two molecules $CH_3CHOHCH_3$ and $CH_3CH_2CH_2OH$ (containing 3 carbons, 8 hydrogens and 1 oxygen) are isomers carrying an alcoholic function on the second or third carbon atom, respectively.

Isotope An atom* is an isotope of another atom if the nuclei of the two atoms contain the same number of protons* but different numbers of neutrons*. For instance, the three carbon isotopes (termed ^{12}C, ^{13}C and ^{14}C) all contain 6 protons and, respectively, 6, 7 and 8 neutrons.

Leaf A plant is said to have opposite leaves when the leaves arise in pairs in which each of the two leaves of the pair is arranged opposite to the other, symmetrically relative to the stem. A plant is said to have alternate leaves when the leaves are at a distance from one another along the stem. Whether the leaves are opposite or alternate, their flattened part is called the limb. The petiole attaches the limb to the stem.

Limb See leaf*.

Lipid Fatty substance. For example, animal and plant fats, oil and butter are mostly made up of lipids. The membranes of living cells are made of a double sheet of lipids associated with proteins.

Macromolecule Molecule made of a very large number of atoms.

Meristem Clump of small undifferentiated cells. Meristems may remain quiescent during some time. In an active meristem, the cells proliferate and progressively differentiate, which will permit the formation and growth of the plant organs (roots, buds, shoots).

Mimosa Plant living spontaneously in tropical areas but easily grown under temperate climate. The leaflets and leaves of *Mimosa* rapidly fold following plant stimulation and then progressively come back to their initial position.

Mineral nutrition The mineral nutrition of a plant is the absorption by the roots of minerals (nitrates, sulphates, phosphates and salts of potassium, calcium and magnesium, trace elements, etc.) that the plant needs for its development.

Mitochondria A cell* organelle performing respiration.

Mitosis Process through which an initial somatic cell (often called "mother" cell) divides into two identical "daughter" cells. Another type of cell division (meiosis), which is not described in this book, is involved in the formation of reproductive cells.

Monochromatic Monochromatic light is ideally a light with a single colour, i.e. a light made of electro-magnetic vibrations with a single frequency. Since it is not possible in practise to select a light with a single frequency, a monochromatic light corresponds in fact to a narrow band of frequencies.

Monocotyledons Higher plants possessing a single cotyledon in their seeds. Among the plants under consideration in this book, Poaceae (formerly called Graminaceae) are monocotyledons.

Monoecious Contrary of dioecious*. Most plants are monoecious, i.e. possessing male and female organs in the same plant.

Monovalent ion Ion* carrying one elementary electrical charge*. For instance, Cl^- is a monovalent anion*.

Morphogenesis Process through which an organism (here a plant) acquires its shape through the formation and growth of its organs in the course of its development.

Morphogenetic A morphogenetic substance is a substance controlling plant morphogenesis*.

Neuron Nerve cell.

Neutron Elementary particle that is electrically neutral (no electrical charge) and the mass of which is approximately 1840 times that of an electron*. The neutron is one of the elementary particles constituting atom* nuclei*.

Nucleolus A diffuse structure within the nucleus* of living cells*. It is in the nucleolus that the mRNA (see RNA*) resulting from DNA transcription* is packaged with proteins to form ribosomes*.

Nucleon Either of the two sorts of particle, proton* and neutron*, constituting the cell nucleus*. The mass of the proton is extremely close to that of the neutron. Therefore, the mass of an atom nucleus may be represented by the total number of nucleons (termed the "mass number") that this atom nucleus contains.

Nucleus (atom nucleus) An atom* is made of a nucleus (responsible for the largest part of the atom mass) and of electrons orbiting the nucleus. The nucleus is made of two particles, protons* (carrying an elementary positive charge) and neutrons* (electrically neutral).

Nucleus (cell nucleus) In the cytoplasm*, the nucleus carries the largest part of the genetic information of the cell*. For detail, see Sects. A.2.2, A.2.3 and A.2.4.

Nuclide Atom nucleus characterized by the number of protons* (termed the charge number) and the number of nucleons* (termed the mass number) that it contains. A nuclide is represented conventionally by indicating the charge number in the lower left and the mass number in the upper left of the chemical symbol of the atom under consideration. For example, the nuclide ${}^{12}{}_{6}C$ corresponds to the carbon isotope* the nucleus of which contains 6 protons and 12 nucleons. Since the charge number characterizes the chemical nature of the atom under consideration, the charge number is often omitted (e.g. ${}^{12}C$ instead of ${}^{12}{}_{6}C$). In similar manner the nuclide ${}^{31}{}_{15}P$ (or ${}^{31}P$) corresponds to the most common isotope of phosphorus (containing 15 protons and 31 nucleons).

Nutrient medium Synonymous for nutrient solution*.

Nutrient solution Under natural conditions, plants grow spontaneously on the ground where they can, through their roots, find the water and the minerals necessary for their nutrition. However, under laboratory conditions, it is often preferred to cultivate them in an aqueous solution, the composition of which is controlled so that it provides the plants with the salts and oligo-elements that they need to grow.

Opposite See leaf*.

Oxymoron Expression that contains an internal contradiction. For people believing that it is impossible that a plant has memory, the expression "plant memory" is an oxymoron.

Parenchyma Internal plant tissue. For instance, the leaf parenchyma is made of rounded cells that contain chlorophyll* and perform photosynthesis*.

Petiole See leaf*.

Pheromones Volatile substances of various chemical natures, emitted at extremely low levels in the atmosphere by living beings (including plants), and which may cause alterations in the behaviour of animals and plants that are susceptible to them.

Phloem Plant tissue that conveys sap from leaves to the rest of the plant. Phloem is also thought to play a role in transferring electrical depolarization signals throughout the plant.

Photoperiodicity Relative lengths of night and day.

Photosynthesis Mechanism by which green plants release oxygen and produce sugars from carbon dioxide and water using the energy they get from the absorption of light.

Phytochrome Molecule made of the association of a protein with a pigment. Two isomeric* forms of phytochrome exist, which we are calling here phytochrome1 and phytochrome2. Phytochrome passes in the form phytochrome1 when it is irradiated by far-red light (i.e. a light close to infra-red) and as phytochrome2 when it is irradiated by a clear red light (close to orange). Under natural light, the two forms of phytochrome coexist, and it is the phytochrome2/phytochrome1 ratio that controls the sensitivity of plants to the relative duration of day and night.

Phytoplankton Entirety of the small-size photosynthetic organisms living in suspension in water. Freshwater and marine phytoplankton both exist.

Plasma membrane The membrane that separates cytoplasm* from the cell wall*.

Polyribosome (or polysome) An mRNA* molecule on which are thread several ribosomes*. Each of these ribosomes is in the process of synthesizing a protein* on the pattern of the mRNA molecule, according to the genetic code*.

Protein Macromolecule made of one or several sequences of amino acids*. The binding of each amino acid with the next in the sequence occurs with loss of a water molecule.

Proteome The entirety of the protein equipment of a cell, of an organism, etc.

Proton Elementary particle that carries one positive elementary electrical charge* and the mass of which is close to that of a neutron*. The proton is one of the elementary particles constituting atom* nuclei*.

Pulvinus Motive bulge at the base of the *Mimosa* petioles*. When a *Mimosa* plant is stimulated (for instance, by touching a leaf), this induces a modification in the swelling of the pulvinus cells, which, in turn, drives the movement of the leaflets and leaves*.

Quiescent A quiescent bud is an inactive bud.

Replicate, replication A cell is said to replicate (or duplicate) its DNA when it makes a copy of its DNA. As a consequence, the cell then possesses two copies of its DNA.

Ribosome Small cell organelle, made of proteins* and rRNA (see RNA*) molecules, which is involved in protein biosynthesis.

RNA Ribonucleic acid. There are several sorts of RNA, such as mRNA (messenger RNA), rRNA (ribosome RNA), sRNA (soluble RNA), etc. For more detail, see the Sects. A.2.5 and A.2.6.

Sap There are two kinds of sap, the "xylem sap" and the "phloem sap". The xylem sap, which consists of water and minerals absorbed by the root, is passively transported from the roots to the leaves by the wood vessels (which are dead xylem* cells forming actual tubes). The phloem sap is formed by enriching, within the leaves, the xylem sap with organic substances originating from photosynthesis. The phloem sap is actively transported towards all the parts of the plant by living phloem* cells.

Sensitization Process by which the repetition of a given, harmful stimulus causes the intensity of the response to another occurrence of the same stimulus to be increased.

Statocytes Special cells, located at the root and shoot tips, which are thought to be involved in plant sensitivity to gravity.

Statoliths Small starch particles present in the statocytes*. The statoliths tend to settle in the direction of gravity and they are thought to be the sensors of gravity within the statocytes*.

Stimulation Synonym for stimulus*.

Stimulus Any sort of event that may induce a reaction in a plant.

Stomata. Pores in the plant epidermis that open and close, under the effect of two kidney-shaped cells, and thus allow/prevent gas exchanges and water evaporation.

Substrate When a protein-catalysed reaction or a transport process occur, the substrate is the substance involved (i.e. transformed or transported) in this process.

Symbiont Each of the two partners engaged in a symbiosis*.

Symbiosis Long-lasting association between two organisms that is beneficial to both partners. For example, lichen is a symbiosis of a fungus with an alga.

Tissue culture A plant tissue culture is a clump of more or less dedifferentiated cells obtained by culture under aseptic conditions from an initial plant fragment.

Tonoplast Membrane that separates cytoplasm* from vacuole(s)* in plant cells.

Translation The translation of an mRNA* consists in producing a protein on the pattern of this mRNA. For more detail, see Sect. A.2.6.

Transcription The transcription of a gene consists in forming an mRNA on the pattern of the gene DNA. Transcription occurs in the nucleus of the cell. Then the mRNA migrates to the cytoplasm. Transcription thus permits to transfer the information brought by the genes from the nucleus to the cytoplasm. For more detail, see Sect. A.2.6.

Transduce (to), transduction The transduction of a stimulus by a plant is the entirety of the sequence of processes within the plant that occur between the perception of the stimulus and the ultimate response of the plant to this stimulus.

Transgene Exogenous gene* inserted into the genome of an organism.

Transgenesis Process through which exogenous genes* are inserted into the genome of an organism.

Triplet (or codon*) Sequence of three nucleotides in a DNA or RNA strand. Using the classical symbols representing the nucleotides (Sects. A.2.3 and A.2.5), AGU is the RNA triplet corresponding to the DNA triplet TCA (Table A2.1).

Trivalent ion Ion* carrying three elementary electrical charges*. For instance, PO_4^{3-} is a trivalent anion*.

Vacuole Subcellular structure containing a mixture of water, mineral ions and other small molecules.

Venus flytrap (or *Dionaea*) Carnivorous plant that captures flies or other small insects using leaf traps closing on the preys.

Wall Envelop made of cellulose and pectin substances that surrounds plant cells.

Xylem Woody plant tissue composed of tubes through which sap circulates from roots to leaves.

Zygote Fertilized ovum.

References

Alberts B, Bray D, Lewis J, Raff M, Roberts K, Watson JD (1983) Molecular biology of the cell. Garland Publishing Inc., New York, 1146 p

Allard HA (1947) Photoperiodic behaviour of Maryland Mammoth tobacco under localization treatments and in grafts with Connecticut broad leaf. J Agric Res 74:15–31

Allard HA, Gabner WW (1941) Responses of some plants to equal and unequal ratios of light and darkness in cycles ranging from 1 hour to 72 hours. J Agric Res 63:305–330

Allmann S, Baldwin IT (2010) Insects betray themselves in nature to predators by rapid isomerization of green leaf volatiles. Science 329:1075–1078

Bailey C, Chen M (1983) Morphological basis of long-term habituation and sensitization in Aplysia. Science 220:91–93

Baroin Tourancheau A, Tsao N, Klobutcher LA, Pearlman RE, Adoutte A (1995) Genetic code deviations in the ciliates: evidence for multiple and independent events. EMBO J 14(13):3262–3267

Beyl CA, Mitchell CA (1977) Automated mechanical stress: application for height control of greenhouse chrysanthemum. HortSci 12:575–577

Bourgeade P, Boyer N, De Jaegher G, Gaspar T (1989) Carry-over of thigmomorphogenetic characteristics in calli derived from *Bryonia dioica* internodes. Plant Cell Tissue Organ Cult 19:199–211

Boyer N, Gaspar T, Lamand M (1979) Modification des isoperoxydases et de l'allongement des entre-noeuds de bryone à la suite d'irritations mécaniques. Z Pflanzenphysiol 93:459–470

Braam J, Davis RW (1990) Rain-, wind- and touch-induced expression of calmodulin and calmodulin-related genes in *Arabidopsis*. Cell 60:357–364

Davies E, Schuster A (1981) Intercellular communication in plants: evidence for a rapidly generated, bidirectionally transmitted wound signal. Proc Natl Acad Sci U S A 78:2422–2426

Davies E, Stankovic B, Vian A, Wood AJ (2012) Where has all the message gone? Plant Sci 185–186:23–32

Desbiez M-O, Kergosien Y, Champagnat P, Thellier M (1984) Memorization and delayed expression of regulatory messages in plants. Planta 160:392–399

Desbiez M-O, Gaspar T, Crouzillat D, Frachisse J-M, Thellier M (1987) Effect of cotyledonary prickings on growth, ethylene metabolism and peroxidase activity in *Bidens pilosus*. Plant Physiol Biochem 25:137–143

Desbiez M-O, Tort M, Thellier M (1991) Control of a symmetry-breaking process in the course of the morphogenesis of plantlets of *Bidens pilosa* L. Planta 184:397–402

Desbiez M-O, Tort M, Monnier C, Thellier M (1998) Asymmetrical triggering of the cell cycle in opposite buds of a young plant, after a slight cotyledonary wound. CR Acad Sci Paris (Sciences de la Vie/Life Sciences) 321:403–407

M. Thellier, *Plant Responses to Environmental Stimuli*,
DOI 10.1007/978-94-024-1047-1

Dicke M, van Loon JJA (2000) Multitrophic effects of herbivore-induced plant volatiles in an evolutionary context. Entomol Exp Appl 97:237–249

Dolmetsch RE, Lewis RS, Goodnow CC, Healy JJ (1997) Differential activation of transcription factors induced by Ca^{2+} response amplitude and duration. Nature 386:855–858

Dudai Y (2004) The neurobiology of consolidations, or, how stable is the engram? Annu Rev Psychol 55:51–86

Falkner R, Falkner G (2003) Distinct adaptability during phosphate uptake by the Cyanobacterium *Anabaena variabilis* reflects information processing about preceding phosphate supply. J Trace Microprobe Tech 21:363–375

Farineau J, Morot-Gaudry JF (2011) La Photosynthèse, Quæ. Versailles, 412 p

Frachisse J-M, Desbiez M-O, Champagnat P, Thellier M (1985) Transmission of a traumatic signal via a wave of electric depolarization and induction of correlation between the cotyledonary buds in *Bidens pilosus*. Physiol Plant 64:48–52

Gagliano M, Renton M, Depczynski M, Mancuso S (2014) Experience teaches plants to learn faster and forget slower in environments where it matters. Oecologia 175:63–72

Gaiser JC, Lomax TL (1991) The gravitropic tomato mutant lazy-2 is altered in signal transduction. Curr Plant Sci Biotechnol Agric 13:928–937

Gamow G (1955) M. Tompkins s'explore lui-même. Dunod, Paris, 286 p

Gautheret RJ (1935) Recherches sur la culture des tissus végétaux. Rev Cytol Cytophysiol Veg 1:1–279

Gayler S (2010) Modélisation de l'effet de facteurs de l'environnement sur la répartition des ressources dans un système végétal mixte. CR Acad Agric Fr 96:89–90

Gayler S, Grams TEE, Kozovits A, Luedemann G, Winkler JB, Priesack E (2006) Analysis of competition effects in mono- and mixed cultures of juvenile beech and spruce by means of the plant growth simulation model PLATHO. Plant Biol 8:503–514

Gayler S, Grams TEE, Heller W, Treutter D, Priesack E (2008) A dynamic model of environmental effects on allocation to carbon-based secondary compounds in juvenile trees. Ann Bot 101:1089–1098

Gehring M, Bubb KL, Henikoff S (2009) Extensive demethylation of repetitive elements during seed development underlies gene imprinting. Science 324:1447–1451

Grémiaux A, Girard S, Guérin V, Lothier J, Baluska F, Davies E, Bonnet P, Vian A (2016) Low-amplitude, high-frequency electromagnetic field exposure causes delay and reduced growth in *Rosa hybrida*. J Plant Physiol 190:44–53

Hammer PA, Mitchell CA, Weiler TC (1974) Height control in greenhouse chrysanthemum by mechanical stress. HortSci 9:474–475

Heil M, Fiala B, Linsenmair E, Zotz G, Menke P (1997) Food body production in *Macaranga triloba* (Euphorbiaceae). A plant investment in anti-herbivore defence via symbiotic ant partners. J Ecol 85:847–861

Henry-Vian C, Vian A, Dietrich A, Ledoigt G, Desbiez M-O (1995) Changes in the polysomal mRNA population upon wound signal expression or storage in *Bidens pilosa*. Plant Physiol Biochem 33:337–344

Herms DA, Mattson WJ (1992) The dilemma of plants: to grow or defend. Q Rev Biol 67:283–335

Hütt MT, Lüttge U, Thellier M (2015) Noise-induced phenomena and complex rhythms: a test scenario for plant systems biology. In: Mancuso S, Shabala S (eds) Rhythms in plants, Chapter 11. Springer, Cham, pp 279–321

Karban R, Baldwin IT (1997) Induced responses to herbivores. The University of Chicago Press, Chicago, 330 p

Karban C, Baldwin IT, Baxter KJ, Laue G, Felton GW (2000) Communication between plants: induced resistance of wild tobacco plants following clipping of neighbouring sagebrush. Oecologia 125:66–71

Knight MR, Campbell AK, Smith SM, Trewavas AJ (1991) Transgenic plant aequorin reports the effect of touch and cold-shock and elicitors on cytoplasmic calcium. Nature 352:524–526

Knight MR, Smith SM, Trewavas AJ (1992) Wind-induced plant motion immediately increases cytosolic calcium. Proc Natl Acad Sci U S A 89:4967–4971

Knight H, Brandt S, Knight MR (1998) A history of stress alters drought calcium signalling pathways in *Arabidopsis*. Plant J 16:681–687

Lefèvre J, Boyer N, Gaspar T, Desbiez M-O (1993) Change of peroxidase activity related to storage of information and immediate or delayed expression of the effect of cotyledon pricking on hypocotyl growth in tomato. In: Welinder KG, Rasmussen SK, Penel C, Greppin H (eds) Plant peroxidases: biochemistry and physiology. Publications de l'Université, Genève, pp 345–348

Lesburguères E, Gobbo OL, Alaux-Cantin S, Hambucken A, Trifilieff P, Bontempi B (2011) Early tagging of cortical networks is required for the formation of enduring associative memory. Science 331:924–928

Lodish H, Berk A, Zipursky SL, Matsudaira P, Baltimore D, Darnell J (2000) Learning and memory. In: Molecular cell biology, 4th edn. W.H. Freeman, New York, section 21–7

Lüttge U, Thellier M (2016) Roles of memory and circadian clock in the ecophysiological performances of plants. Prog Bot 77:73–104

Manning GS (1969) Limiting laws and counterion condensation in polyelectrolyte solutions. I. Colligative properties. J Chem Phys 51:924–933

Matyssek R, Koricheva J, Schnyder H, Ernst D, Munch JC, Osswald W, Pretzsch H (2012) The balance between resource sequestration and retention: a challenge in plant science. In: Matyssek R, Schnyder H, Osswald W, Ernst D, Munch JC, Pretzsch H (eds) Growth and defence in plants – resource allocation at multiple scales, Ecological studies, vol 220. Springer, New York, pp 3–24

McAinsh MR, Hetherington AM (1998) Encoding specificity in Ca^{2+} signalling systems. Trends Plant Sci 3:32–36

Mewis I, Appel HM, Hom A, Raina R, Schultz JC (2005) Major signaling pathways modulate Arabidopsis glucosinolate accumulation and response to both phloem-feeding and chewing insects. Plant Physiol 138:1149–1162

Nick P, Schäfer E (1988) Spatial memory during the tropism of maize. Planta 175:380–388

Pandolfi C, Azzarello E, Mancuso S, Shabala S (2016) Acclimation improves salt stress tolerance in *Zea mays* plants. J Plant Physiol 201:1–8

Plieth C, Hansen UP, Knight H, Knight MR (1999) Temperature sensing by plants: the primary characteristics of signal perception and calcium response. Plant J 18:491–497

Pontier D, Balague C, Roby D (1998) The hypersensitive response. A programmed cell death associated with plant resistance. CR Acad Sci Paris (série III) 321:731–734

Roberts JA (1984) Tropic responses of hypocotyls from normal tomato plants and the gravitropic mutant *Lazy-1*. Plant Cell Environ 7:515–520

Roux D, Vian A, Girard S, Bonnet P, Paladian F, Davies E, Ledoigt G (2006) Electromagnetic fields (900 MHz) evoke consistent molecular responses in tomato plants. Physiol Plant 128:283–288

Schrödinger E (1944) What is life? The physical aspect of the living cell. University Printing House, Cambridge, 96 p

Shepherd VA (2005) From semi-conductors to the rhythm of sensitive plants: the research of J.C. Bose. Cell Mol Biol 51:607–619

Tafforeau M, Verdus M-C, Charlionet R, Cabin-Flaman A, Ripoll C (2002a) Two-dimensional electrophoresis investigation of short-term response of flax seedlings to cold shock. Electrophoresis 23:2534–2540

Tafforeau M, Verdus M-C, Norris V, White G, Demarty M, Thellier M, Ripoll C (2002b) SIMS study of the calcium-deprivation step related to epidermal meristem production induced in flax by cold shock or radiation from a GSM telephone. J Trace Microprobe Tech 20:611–623

Tafforeau M, Verdus M-C, Norris V, White GJ, Cole M, Demarty M, Thellier M, Ripoll C (2004) Plant sensitivity to low intensity 105 GHz electromagnetic radiation. Bioelectromagnetics 25:403–407

Tafforeau M, Verdus M-C, Norris V, Ripoll C, Thellier M (2006) Memory processes in the response of plants to environmental signals. Plant Signal Behav 1:9–14

Thellier M (2015) Les plantes ont-elles une mémoire? In: Legrand M (ed), Quæ, Versailles, 111 p

Thellier M, Lüttge U (2013) Plant memory. Plant Biol 15:1–12

Thellier M, Desbiez M-O, Kergosien Y, Champagnat P (1981) Mise en mémoire de signaux morphogènes chez *Bidens pilosus* L. CR Acad Sci Paris (série III) 292:1187–1190

Thellier M, Desbiez M-O, Champagnat P, Kergosien Y (1982) Do memory processes occur also in plants ? Physiol Plant 56:281–284

Thellier M, Ripoll C, Norris V (2013) Memory processes in the control of plant growth and metabolism. Nova Acta Leopoldina NF 114(391):21–42

Trewavas A (1999) Le calcium c'est la vie: calcium waves. Plant Physiol 120:1–6

Trewavas A (2003) Aspects of plant intelligence. Ann Bot 92:1–20 (see the section "Plant memory and information retrieval")

Ueda M, Nakamura Y (2006) Metabolites involved in plant movement and "memory": nyctinasty of legumes and trap movement in the Venus flytrap. Nat Prod Rep 23:548–557

Verdus M-C, Cabin-Flaman A, Ripoll C, Thellier M (1996) Calcium-dependent storage/retrieval of environmental signals in plant development. CR Acad Sci Paris (Sciences de la Vie/Life Sciences) 319:779–782

Verdus M-C, Thellier M, Ripoll C (1997) Storage of environmental signals in flax: their morphogenetic effect as enabled by a transient depletion of calcium. Plant J 12:1399–1410

Verdus M-C, Le Sceller L, Norris V, Thellier M, Ripoll C (2007) Pharmacological evidence for calcium involvement in the long-term processing of abiotic stimuli in plants. Plant Signal Behav 2:212–220

Verdus M-C, Ripoll C, Norris V, Thellier M (2012) The role of calcium in the recall of stored morphogenetic information by plants. Acta Biotheor 60:83–97

Vian A, Roux D, Girard S, Bonnet P, Paladian F, Davies E, Ledoigt G (2006) Microwave irradiation affects gene expression in plants. Plant Signal Behav 1:67–70

Volkov AG, Carrel H, Adesina T, Markin VS, Jovanov E (2008) Plant electrical memory. Plant Signal Behav 3:490–492

Wheeler RM, Salisbury FB (1979) Water spray as a convenient means of imparting mechanical stimulation to plants. HortSci 14:270–271

Young TP, Stubblefield CH, Isbell LA (1997) Ants on swollen-thorn acacias: species coexistence in a simple system. Oecologia 109:98–107

MIX
Papier aus verantwortungsvollen Quellen
Paper from responsible sources
FSC® C105338

If you have any concerns about our products,
you can contact us on
ProductSafety@springernature.com

In case Publisher is established outside the EU,
the EU authorized representative is:
Springer Nature Customer Service Center GmbH
Europaplatz 3, 69115 Heidelberg, Germany

Printed by Libri Plureos GmbH
in Hamburg, Germany